Springer-Science+Business Media, B.V.

Dear Reader

We would very much appreciate receiving your suggestions and criticisms for the *Plant Molecular Biology Manual.* They will prove to be most helpful during our preparations for future supplements.

Would you please answer the questions listed below, and send your comments with any further suggestions you may have, to *Drs. Gilles Jonker* at the above-mentioned address.

Thank you for your assistance!

Drs. Gilles Jonker
Publisher

AF763431

PLANT MOLECULAR BIOLOGY MANUAL

1. What errors have you found? (list page numbers and describe mistakes)
2. What protocols do you find to be confusing or lacking in detail? (list chapter numbers and page numbers and describe problems)
3. What protocols do you feel should be replaced in future supplements with newer (better) methods?
4. What new topics or other material would you like to see included in future supplements?

Please print or type your answers in the space below, and continue overleaf.

Name: Date:

Address:

Plant Molecular Biology Manual
Supplement 4, August 1991

INSTRUCTIONS FOR SUPPLEMENT 4:

Preliminary pages
All preliminary pages should be replaced by new pages i–xviii.

Section B
Add after Chapter B15:

Chapter B16: J.L. Slightom *et al.*/Custom polymerase chain reaction engineering plant expression vectors and genes for plant expression

Section C
Add after Chapter C2:

Chapter C3: P.P.Chee *et al.*/Using polymerase chain reaction to identify transgenic plants

Index
Replace

PLANT MOLECULAR BIOLOGY MANUAL

PLANT MOLECULAR BIOLOGY MANUAL

Edited by

STANTON B. GELVIN
Purdue University, West Lafayette
Indiana, USA

ROBBERT A. SCHILPEROORT
Leiden State University, Leiden
The Netherlands

DESH PAL S. VERMA
Ohio State University, Columbus
Ohio, USA

Springer-Science+Business Media, B.V.

Library of Congress Catalog Card Number: 88-198312

ISBN 978-0-7923-1341-0 ISBN 978-94-017-4820-9 (eBook)
DOI 10.1007/978-94-017-4820-9

Printed on acid-free paper

Contents

SECTION A: Introduction of DNA into Plant Cells

1. Direct DNA transfer to protoplasts with and without electroporation
   Michael W. Saul, Raymond D. Shillito & Ioan Negrutiu

2. Use of cointegrating Ti plasmid vectors
   Stephen G. Rogers, Harry Klee, Robert B. Horsch & Robert T. Fraley

3. Binary vectors
   Gynheung An, Paul R. Ebert, Amitava Mitra & Sam B. Ha

4. *Agrobacterium* molecular genetics
   Paul J.J. Hooykaas

5. Leaf disc transformation
   Robert B. Horsch, Joyce Fry, Nancy Hoffmann, Jeanne Neidermeyer, Stephen G. Rogers & Robert T. Fraley

6. Extraction of DNA from plant tissues
   Scott O. Rogers & Arnold J. Bendich

7. Procedures for constructing ds-cDNA clone banks
   Jerry L. Slightom & Hector D. Quemada

8. Procedures for constructing genomic clone banks
   Jerry L. Slightom & Roger F. Drong

9. Selectable and screenable markers
   Arlette Reynaerts, Marc de Block, Jean-Pierre Hernalsteens & Marc van Montagu

10. Transformation of plant pathogenic fungi
    Robert C. Garber, B. Gillian Turgeon & Olen C. Yoder

*11. Isolation of mitochondria and mtDNA
    Hanna Skubatz & Arnold J. Bendich

---

* Included in Supplement 2.

*12. DNA transfer into cyanobacteria
Jeff Elhai, Teresa Thiel & Himadri B. Pakrasi

*13. Direct DNA transfer into intact plant cells and recovery of transgenic plants via microprojectile bombardment
D.T. Tomes, M. Ross, R. Higgens, A.G. Rao, M. Staebell & J. Howard

*14. Isolation of plant genes by heterologous complementation in *Escherichia coli*
Ashton J. Delauney & Desh Pal S. Verma

†15. Preparation of high molecular weight plant DNA and analysis by pulsed-field gel electrophoresis
Raymond A.J. van Daelen & Pim Zabel

SECTION B: Expression of Genes in Plants

1. Use of reporter genes to study gene expression in plant cells
Luis Herrera-Estrella, Teemu H. Teeri & June Simpson

2. Assays for studying chromatin structure
Anna-Lisa Paul & Robert J. Ferl

3. Assays for studying DNA methylation
Aharon Razin

4. Analysis of gene expression in transgenic plants
Ferenc Nagy, Steve A. Kay & Nam-Hai Chua

5. Subcellular targeting of proteins *in vivo* and *in vitro*
P.H. Schreier, B. Reiss & M. Kuntz

6. Isolation of total and polysomal RNA from plant tissues
Sacco de Vries, Harry Hoge & Ton Bisseling

7. Translation in *Xenopus* oocytes of mRNAs transcribed *in vitro*
E.E. Kawata, G. Galili, L.D. Smith & B.A. Larkins

---

* Included in Supplement 2.
† Included in Supplement 3.

8. Use of the firefly luciferase gene as a reporter of gene expression in plants
Stephen H. Howell, David W. Ow & Michel Schneider

9. *In situ* RNA hybridization in plant tissues
Natasha V. Raikhel, Y. Bednarek & David R. Lerner

10. Light-regulated promoter sequences in the multigene *Cab* and *rbcS* gene families
Robert Fluhr

11. *In vitro* DNA footprinting
Pamela J. Green, Steve A. Kay, Eric Lam & Nam-Hai Chua

12. *In vivo* detection of protein-DNA interactions
Robert J. Ferl

*13. Electron microscopic immunocytochemistry in plant molecular biology
Eliot M. Herman & Diane L. Melroy

†14. The GUS gene fusion system
Richard A. Jefferson & Kate J. Wilson

†15. The use of *lacZ* as a reporter gene in plants
Teemu H. Teeri & Luis Herrera-Estrella

§16. Custom polymerase chain reaction engineering plant expression vectors and genes for plant expression
Jerry L. Slightom, Roger F. Drong, Leang C. Sieu & Paula P. Chee

SECTION C: Fate of Introduced Genes

1. Stability of introduced genes and stability in expression
Pamela Dunsmuir, Diane Bond, Kathleen Lee, David Gidoni & Jeffrey Townsend

2. Restriction fragment length polymorphism
Robert Bernatzky

§3. Using polymerase chain reaction to identify transgenic plants
Paula P. Chee, Roger F. Drong & Jerry L. Slightom

INDEX §

* Included in Supplement 2.
† Included in Supplement 3.
§ Included in Supplement 4.

List of Contributors

Author	*Chapter*	*Address*
An, G.	A3	Institute of Biological Chemistry and Plant Physiology Program, Washington State University, Pullman, WA 99164-6340, USA
Bednarek, S.Y.	B9	MSU-DOE Plant Research Laboratory, Michigan State University, East Lansing, MI 48824-1312, USA
Bendich, A.J.	A6, A11	Departments of Botany and Genetics, University of Washington, Seattle, WA 98195, USA
Bernatzky, R.	C2	Department of Plant and Soil Science, University of Massachusetts, Amherst, MA 01003, USA
Bisseling, T.	B6	Department of Molecular Biology, Agricultural University, De Dreijen 11, 6703 BC Wageningen, The Netherlands
Bond, D.	C1	Advanced Genetic Sciences, 6701 San Pablo Avenue, Oakland, CA 94608, USA
Chee, P.P.	B16, C3	Division of Molecular Biology, The Upjohn Company, Kalamazoo, MI 49007, USA
Chua, N.-H.	B4, B11	Laboratory of Plant Molecular Biology, The Rockefeller University, 1230 York Avenue, New York, NY 10021-6399, USA
De Block, M.	A9	Plant Genetic Systems N.V., J. Plateaustraat 22, B-9000 Ghent, Belgium
Delauney, A.J.	A14	Department of Molecular Genetics and The Ohio State Biotechnology Center, The Ohio State University, Columbus, OH 43210, USA

Author	*Chapter*	*Address*
De Vries, S.	B6	Department of Molecular Biology, Agricultural University, De Dreijen 11, 6703 BC Wageningen, The Netherlands
Drong, R.F.	A8, B16, C3	Division of Molecular Biology, The Upjohn Company, Kalamazoo, MI 49007, USA
Dunsmuir, P.	C1	Advanced Genetic Sciences, 6701 San Pablo Avenue, Oakland, CA 94608, USA
Ebert, P.R.	A3	Institute of Biological Chemistry and Plant Physiology Program, Washington State University, Pullman, WA 99164-6340, USA
Elhai, J.	A12	MSU-DOE Plant Research Laboratory, Michigan State University, E. Lansing, MI 48824, USA
Ferl, R.J.	B2, B12	Department of Vegetable Crops, University of Florida, Gainesville, FL 32611, USA
Fluhr, R.	B10	Department of Plant Genetics, Weizmann Institute of Science, Rehovot 76100, Israel
Fraley, R.T.	A2, A5	Plant Molecular Biology, Corporate Research Laboratories, Monsanto Company, St. Louis, MO 63198, USA
Fry, J.	A5	Plant Molecular Biology, Corporate Research Laboratories, Monsanto Company, St. Louis, MO 63198, USA
Galili, G.	B7	Department of Botany and Plant Pathology, Purdue University, Lilly Hall of Life Sciences, West Lafayette, IN 47907, USA
Garber, R.C.	A10	Biotechnology Center and Department of Plant Pathology, Ohio State University, Columbus, OH 43210, USA

Author	*Chapter*	*Address*
Gidoni, D.	C1	Advanced Genetic Sciences, 6701 San Pablo Avenue, Oakland, CA 94608, USA
Green, P.J.	B11	Laboratory of Plant Molecular Biology, The Rockefeller University, 1230 York Avenue, New York, NY 10021-6399, USA
Ha, S.B.	A3	Institute of Biological Chemistry and Plant Physiology Program, Washington State University, Pullman, WA 99164-6340, USA
Herman, E.M.	B13	Plant Molecular Biology Laboratory, Bldg. 006, USDA, Agricultural Research Service, Beltsville, MD 20705, USA
Hernalsteens, J.-P.	A9	Laboratorium voor Genetische Virologie, Vrije Universiteit Brussel, Paardenstraat 65, B-1640 St. Genesius-Rode, Belgium
Herrera-Estrella, L.	B1, B15	Centro de Investigación y de Estudios Avanzados del I.P.N., Unidad Irapuato, 36500 Irapuato, Gto., Mexico
Higgens, R.	A13	Pioneer Hi-Bred International, Inc., Department of Biotechnology Research, Johnston, IA 50131-0038, USA
Hoffman, N.	A5	Plant Molecular Biology, Corporate Research Laboratories, Monsanto Company, St. Louis, MO 63198, USA
Hoge, H.	B6	Department of Biochemistry, State University of Leiden, Wassenaarseweg 64, 2333 AL Leiden, The Netherlands
Hooykaas, P.J.J.	A4	Department of Plant Molecular Biology, Biochemistry Laboratory, State University of Leiden, Wassenaarseweg 64, 2333 AL Leiden, The Netherlands

Author	*Chapter*	*Address*
Horsch, R.B.	A2, A5	Plant Molecular Biology, Corporate Research Laboratories, Monsanto Company,
Howard, J.	A13	Pioneer Hi-Bred International, Inc., Department of Biotechnology Research, Johnston, IA 50131-0038, USA
Howell, S.H.	B8	Boyce Thompson Institute at Cornell University, Tower Road, Ithaca, NY 14853-1801, USA
Jefferson, R.A.	B14	Food and Agriculture Organization of the United Nations and the International Atomic Energy Agency, Wagramerstrasse 5, P.O. Box 200, A-1400 Vienna, Austria
Kawata, E.E.	B7	Department of Botany and Plant Pathology, Purdue University, Lilly Hall of Life Sciences, West Lafayette, IN 47907, USA
Kay, S.A.	B4, B11	Laboratory of Plant Molecular Biology, The Rockefeller University, 1230 York Avenue, New York, NY 10021-6399, USA
Klee, H.	A2	Plant Molecular Biology, Corporate Research Laboratories, Monsanto Company, St. Louis, MO 63198, USA
Kuntz, M.	B5	Max Planck Institut, Abt. Prof. J. Schell, D-5000 Cologne 30, Germany
Lam, E.	B11	Laboratory of Plant Molecular Biology, The Rockefeller University, 1230 York Avenue, New York, NY 10021-6399, USA
Larkins, B.A.	B7	Department of Botany and Plant Pathology, Purdue University, Lilly Hall of Life Sciences, West Lafayette, IN 47907, USA
Lee, K.	C1	Advanced Genetic Sciences, 6701 San Pablo Avenue, Oakland, CA 94608, USA

Author	*Chapter*	*Address*
Lerner, D.	B9	MSU-DOE Plant Research Laboratory, Michigan State University, East Lansing, MI 48824-1312, USA
Melroy, D.L.	B13	Plant Molecular Biology Laboratory, Bldg. 006, USDA, Agricultural Research Service, Beltsville, MD 20705, USA
Mitra, A.	A3	Institute of Biological Chemistry and Plant Physiology Program, Washington State University, Pullman, WA 99164-6340, USA
Nagy, F.	B4	Laboratory of Plant Molecular Biology, The Rockefeller University, 1230 York Avenue, New York, NY 10021-6399, USA
Negrutiu, I.	A1	Institute for Molecular Biology, Paardenstraat 65, B-1640 St. Genesius-Rode, Belgium
Neidermeyer, J.	A5	Plant Molecular Biology, Corporate Research Laboratories, Monsanto Company, St. Louis, MO 63198, USA
Ow, D.W.	B8	USDA/UCB Plant Gene Expression Center, 800 Buchanan Street, Albany, CA 94710, USA
Pakrasi, H.B.	A12	Plant Biology Program, Department of Biology, Box 1137, Washington University, St. Louis, MD 63130, USA
Paul, A.-L.	B2	Department of Botany, University of Florida, Gainesville, FL 32611, USA
Quemada, H.D.	A7	Division of Molecular Biology, The Upjohn Company, Kalamazoo, MI 49007, USA
Raikhel, N.V.	B9	MSU-DOE Plant Research Laboratory, Michigan State University, East Lansing, MI 48824-1312, USA

Author	*Chapter*	*Address*
Rao, A.G.	A13	Pioneer Hi-Bred International, Inc., Department of Biotechnology Research, Johnston, IA 50131-0038, USA
Razin, A.	B3	Department of Cellular Biochemistry, The Hebrew University, Hadassah Medical School, Jerusalem, Israel 91010
Reiss, B.	B5	University of Arizona, Dept. of Biochemistry, Tucson, AZ 85721, USA
Reynaerts, A.	A9	Plant Genetic Systems N.V., J. Plateaustraat 22, B-9000 Ghent, Belgium
Rogers, S.G.	A2, A5	Plant Molecular Biology, Corporate Research Laboratories, Monsanto Company, St. Louis, MO 63198, USA
Rogers, S.O.	A6	Department of Botany, KB-15, University of Washington, Seattle, WA 98195, USA
Ross, M.	A13	Pioneer Hi-Bred International, Inc., Department of Biotechnology Research, Johnston, IA 50131-0038, USA
Saul, M.W.	A1	Friedrich Miescher Institute, P.O. Box 2543, CH-4002 Basel, Switzerland
Schreier, P.H.	B5	Bayer AG, PF-A-BF, 6240, D-5090 Leverkusen, Germany
Shillito, R.D.	A1	Ciba-Geigy Corp., P.O. Box 12257, Research Triangle Park, NC 27709-2257, USA
Sieu, L.C.	B16, C3	Division of Molecular Biology, The Upjohn Company, Kalamazoo, MI 49007, USA
Simpson, J.	B1	Centro de Investigación y de Estudios Avanzados del I.P.N., Unidad Irapuato, 36500 Irapuato, Gto., Mexico

Author	*Chapter*	*Address*
Skubatz, H.	A11	Department of Botany, University of Washington, Seattle, WA 98195, USA
Slightom, J.L.	A7, A8, B16, C3	Division of Molecular Biology, The Upjohn Company, Kalamazoo, MI 49007, USA
Smith, L.D.	B7	Department of Biological Sciences, Purdue University, West Lafayette, IN 47907, USA
Staebell, M.	A13	Pioneer Hi-Bred International, Inc., Department of Biotechnology Research, Johnston, IA 50131-0038, USA
Teeri, T.H.	B1, B15	Molecular Genetics Laboratory, Dept. of Genetics, University of Helsinki, Arkadiankatu 7, SF-00100 Helsinki 10, Finland
Thiel, T.	A12	Department of Biology, University of Missouri-St. Louis, St. Louis, MO 63121, USA
Tomes, D.T.	A13	Pioneer Hi-Bred International, Inc., Department of Biotechnology Research, Johnston, IA 50131-0038, USA
Townsend, J.	C1	Advanced Genetic Sciences, 6701 San Pablo Avenue, Oakland, CA 94608, USA
Turgeon, B.G.	A10	Department of Plant Pathology, Cornell University, Ithaca, NY 14853, USA
Van Daelen, R.A.J.	A15	Department of Molecular Biology, Wageningen Agricultural University, Dreyenlaan 3, 6703 HA Wageningen, The Netherlands
Van Montagu, M.	A9	Genetic Laboratory, State University of Ghent, Ledeganckstraat 35, B-9000 Ghent, Belgium

Author	*Chapter*	*Address*
Verma, D.P.S.	A14	Department of Molecular Genetics and The Ohio State Biotechnology Center, The Ohio State University, Columbus, OH 43210, USA
Wilson, K.J.	B14	Department of Biochemistry and Biological Sciences, Wye College, University of London, Wye, Ashford, Kent TN25 5AH, UK
Yoder, O.C.	A10	Department of Plant Pathology, Cornell University, Ithaca, NY 14853, USA
Zabel, P.	A15	Department of Molecular Biology, Wageningen Agricultural University, Dreyenlaan 3, 6703 HA Wageningen, The Netherlands

Preface

During the past ten years, great advances have been made in the area of plant molecular biology. Such formerly esoteric techniques as gene transfer and plant regeneration are now routinely performed, making the dissection of regulatory elements of genes a common practice in many laboratories. Along with this new technology has come an almost bewildering array of rapidly changing techniques, often making it difficult for the novice to select and perform the technique most appropriate for answering a given biological question. In 1986, some of us felt that many of these techniques had become routine enough to warrant the publication of a laboratory manual. The manual is designed both for advanced college level laboratory courses and as a 'bench guide' for use in the scientific laboratory. Recognizing the rapidly changing nature of plant molecular biology technology, the editors have designed a laboratory manual that is both easy to use in the laboratory and which will be updated as the techniques change and new technologies are devised. Additional chapters that can replace or be added to this first edition will be published periodically.

The editors recognize that many of the techniques described in this manual depend upon specialized plant genetic material, microbial strains, or recombinant plasmids. Those people desiring such material should contact the relevant authors directly. A list of the various contributors to this manual, including their addresses, is included.

The editors would like to thank the authors for the speed with which they contributed their chapters. Special thanks go to Ms. Wilma Foust (Purdue University) for her secretarial assistance.

Preface

[illegible]

Plant Molecular Biology Manual **B16**: 1–55, 1991.

Custom polymerase chain reaction engineering plant expression vectors and genes for plant expression

JERRY L. SLIGHTOM, ROGER F. DRONG, LEANG C. SIEU & PAULA P. CHEE
Molecular Biology–Unit 7242, The Upjohn Company, Kalamazoo, MI 49007, USA

Introduction

In Chapter C3 we have shown how the polymerase chain reaction (PCR) amplification can be an extremely useful technique for the identification of transgenic plants. In this chapter we continue to demonstrate the vast capabilities of the PCR technique by showing that it can be used to aid in the construction of plant expression vectors and the engineering of beneficial genes for expression in these vectors. The engineered expression vectors and genes described in this chapter have the potential to produce superior plant varieties that are protected from plant virus infections (Sections 1 and 2) or produce seed proteins with improved nutritional quality (Sections 3 to 5).

Many molecular biology experiments require the modification of genetic elements–coding regions and gene regulatory elements–for expression in heterologous expression systems, or to make genetic alterations that affect the level, tissue location, and/or developmental timing of gene expression in a homologous expression system. Molecular biology techniques for the modification of DNA molecules are numerous and include the use of linkers, DNA-modifying enzymes to change restriction enzyme sites (Klenow, T4 polymerase, S1 nuclease, mung bean nuclease), and single-stranded site-specific modification methods (see [30]). All of these methods are useful, but they have their limitations, or they rely on random events that require a considerable amount of time and effort to locate the modified product. In an attempt to avoid these limitations we have investigated the use of PCR amplification to incorporate modifications into DNA fragments. The use of PCR amplifications to modify a DNA fragment was first discussed by Mullis *et al.* [33], and first demonstrated by Scharf *et al.* [47], who showed that restriction enzyme sites synthesized into the 5′ ends of PCR oligonucleotide primers could be incorporated into the PCR product even though these recognition sequences are mismatched with the initial template. This mismatch only occurs early in the PCR process because after the first few cycles of DNA synthesis, DNA fragments containing the added restriction enzyme sites are the predominant PCR template [16, 21].

Our PCR-based strategy involves the construction of expression cassettes that contain only one internal restriction enzyme cloning site; however, their use can be virtually universal because identical restriction enzyme sites are PCR-engineered onto the targeted coding regions. Because this technique requires the use of custom-synthesized oligonucleotide primers for each DNA modification, it has been referred to as custom-PCR (CPCR) engineering by Slightom [53]. A similar use of PCR engineering of *Escherichia coli* expression cassettes has been described by MacFerrin *et al.* [28].

The major strength of CPCR engineering is that most of the complex DNA modifications can be incorporated into the sequence of the oligonucleotide primers, and in many cases the engineering process can proceed as a series of straightforward ligation events. The method does not require the use of random events such as those required when using *Bal*31 digestion or M13 site-specific modifications. However, a major weakness of the strategy is the concern that *Taq* polymerase is subject to a high nucleotide incorporation error rate [46]. More recent analyses of the error rate of *Taq* polymerase find the rate to be similar to that of other polymerase, i.e., T7 and *E. coli* DNA polymerase I [25, 54]. In most of the examples presented in this chapter, except Section 4 where total genomic DNA is used, the probability of incorporating PCR-generated errors is significantly reduced because the initial template is cloned and its concentration (100 ng or more) is not limiting as is the case when amplifying rare mRNAs or single-copy genomic DNAs. Thus, the probability of PCR-generated errors (that occur early in the amplification process) becoming the predominant PCR amplified DNA species in a reaction is minimal. This assumption has been supported by nucleotide sequence analysis of the CPCR-engineered clones described by Slightom [53] and in this chapter, where a total of 8400 bp of CPCR-engineered DNAs were analysed and no apparent PCR-generated differences were found. However, it was surprising that several errors were found that occurred during ligation events [53]. These ligation errors may have been the result of exonuclease activities present in the restriction enzyme or ligase, or they may have been synthesis errors in the oligonucleotide primer. DNA synthesis errors can be a serious problem and they can only be identified by nucleotide sequencing the oligonucleotide primer [31, 52]. However, none of the oligonucleotide primers used in the examples presented in this chapter were sequenced. We have found that results such as the CPCR reactions not amplifying the correct size fragment or the amplified fragments not being cloned are indicative of potential nucleotide synthesis error(s) in the oligonucleotide primer(s). Thus, although CPCR is a simple and straightforward method for DNA modification, it is not completely error-free, and for this reason it suffers from the additional requirement that the engineered DNAs should be completely sequenced. This is not a severe problem because the use of T7 polymerase has greatly simplified DNA sequencing; we have developed DNA sequencing methods for rapid nucleotide sequence analysis of CPCR inserts as large as 1 kb [48, 52].

In Section 1 we outline the basic CPCR amplification protocol, and we describe CPCR engineering the plant constitutive expression vector referred to as pUC18cpexp. Section 1 is the only section that contains the complete description of all the important steps involved in CPCR engineering. In the later sections we concentrate on the process involved in the selection of oligonucleotide primers to obtain a desired CPCR-engineered product; most of the CPCR steps refer to the protocol described in Section 1. If additional steps are required they are included in the appropriate section. In Section 2, we demonstrate the use of CPCR amplification to modify a cloned plant virus capsid or coat protein gene by adding flanking *Nco* I sites, allowing it to be cloned into the constitutive expression cassette pUC18cpexp. In Sections 3 and 4, we demonstrate the use of CPCR amplification to construct seed-specific expression cassettes; first (Section 3) using the cloned phaseolin seed storage protein gene of *Phaseolus vulgaris* [50], followed by (Section 4) the engineering of the promoter region of the gene encoding the α' subunit of β-conglycinin of *Glycine max* [15] directly from total soybean genomic DNA. This latter example clearly demonstrates the immense power of this technique because any known DNA sequence (gene regulatory and coding regions) can be engineered by anyone who has the genomic DNA. In Section 5, we demonstrate a novel use of this technology to aid in the purification and modification of a very long oligonucleotide. We demonstrate the use of CPCR amplification to purify and modify a synthetic 144 nucleotide oligonucleotide from its highly contaminated end-synthesis mixture. The use of this technique should improve the rate of chemical synthesis of complete genes.

Section 1. Construction of a constitutive plant expression cassette

The regulatory sequence most widely used to control constitutive expression in plant tissues (both monocot and dicot species) is that responsible for the expression of the cauliflower mosaic virus (CaMV) 35S transcript [18]. The nucleotide sequence identity of many of the elements that control transcription of this 35S transcript has been determined [6, 7, 36]. Many expression cassettes have been constructed that fuse the 5′- and 3′-CaMV 35S gene regulatory elements flanking a multi-cloning site. Pietrzak *et al.* [37] describe the construction of this type of a CaMV 35S-based cassette referred to as pDH51. Expression cassettes of this type are valuable and they promote efficient transcription, but because they rely on the very short 35S 5′-untranslated region they sometimes do not promote efficient translation of the fused coding region. Gallie *et al.* [17] demonstrated that the translation efficiency of mRNAs can be greatly increased by the addition of plant virus-derived 5′-untranslated regions. In particular, the 5′-untranslated regions from tobacco mosaic virus [17] and alfalfa mosaic virus [22] appear to be effective.

The 5′-untranslated region of the cucumber mosaic virus (CMV) strain C and its adjacent coat protein have recently been expressed in plants following cloning between the regulatory sequences of the CaMV 35S transcript [41]. The CMV 5′-untranslated region also appears to promote a high level of translation as determined by production of the CMV coat protein in transgenic plants [41]. Engineering the CMV coat protein gene for expression was relatively straightforward because the coat protein gene is naturally expressed as a subgenomic RNA of CMV RNA3. Thus, the use of its 5′-untranslated region (70 nt in length) and translation initiation codon (ATG) [36] was expected to promote efficient translation in plant tissues. The cloning of the CMV 5′-untranslated region and coat protein gene involved 5′ and 3′ transcriptional fusions with the regulatory elements of the CaMV 35S promoter and the clone referred to as pUC1813cpCMV19 was isolated [41]. However, the engineering of gene coding regions of many other genes can be much more difficult if they do not contain a 5′-untranslated region or even the translation initiation codon (ATG), as is the case for the coat protein genes from a group of plant viruses known as potyviruses (see Section 2). These important gene expression regulatory elements must be added as part of the genetic engineering steps.

Genetic engineering processes that involve translational fusions are much more difficult because of the added requirement that the reading frame of the gene must be preserved. We decided to construct an expression cassette in which the translational fusions are made at the translation initiation (ATG) site using a restriction enzyme recognition site that contains the translation initiation codon. The number of restriction enzyme sites that encode the translation initiation codon is limited, but our analysis suggested that the *Nco* I site could be particularly useful. The *Nco* I site is useful because its flanking sequences share considerable identity with the plant consensus translation initiation sequence [26, 27], AAXXATGG, where X can be any nucleotide. In the case of the *Nco* I site the Xs are represented by two C nucleotides. (Also note that the oligonucleotide primers used to engineer the consensus translation initiation site must include the two A nucleotides preceeding the *Nco* I recognition site; see oligonucleotide primer JLS-82 in Fig. 1.)

One important feature designed for all oligonucleotide primers used for engineering restriction enzyme cut sites is the addition of six nucleotides 5′-adjacent to the synthesized site. These nucleotides were found to aid in the digestion of these synthesized sites and thus increased the number of cut fragments available for ligation and increased the number of clones obtained. We found that the addition of at least 6 'extra' nucleotides was extremely important; the addition of only three nucleotides did not promote a significant amount of enzyme cutting (J.L.Slightom, unpublished data). It appears that many restriction enzymes require a minimum of at least four to five nucleotides flanking a CPCR engineered site before any cutting takes place [23].

The following outline presents the general protocol and techniques needed for CPCR engineering. Within this general outline we present the step-by-step results obtained during the engineering of a new CaMV 35S promoter-based expression cassette. This new expression cassette (see Fig. 3) is based on CPCR engineering of the fused CaMV-CMV promoter-5′-untranslated and CaMV terminator regions of pUC1813cpCMV19 (Fig. 1).

1.1. CPCR engineering procedure: construction of the plant expression vector pUC18cpexp

Steps in the procedure

1. Determine the type of modifications that need to be made. From the analysis of the target sequence and the end product, design a route to obtain the objective goal(s) that involves the addition of modified sequences designed within custom synthesized oligonucleotide primers. If more than one cloning vector is required for the gene transfer process

(which is the case for many gene transfers in plants) determine which restriction enzyme sites will be the most useful for the initial and subsequent transfers. It is difficult to outline a procedure that can cover all the intricacies involved in designing a CPCR-modified DNA fragment. It is more informative to provide an example, such as CPCR-engineering the insert of vector pUC18cpexp.

From the sequence of pUC1813cpCMV19 (Fig. 1) two sets of oligonucleotide primers were designed: one set to CPCR-engineer the fused CaMV-CMV promoter-5′-untranslated region, and the other set to CPCR-engineer the CaMV terminator region (Fig. 1). Each primer shares at least 20 nucleotides with the target DNA region to ensure proper annealing. Primer JLS-81 binds 330 bp 5′ of the CaMV translation initiation site and adds a 5′-*Hind*III site (used for cloning into pUC18 and the binary vector pGA482 [4]) and primer JLS-82 binds to the CMV translation initiation codons and modifies its sequence to the *Nco*I recognition sequence. Primers JLS-83 and JLS-84 are designed to amplify a 200 bp region of the CaMV 35S terminator region and add *Nco*I and *Hind*III sites, respectively (Fig. 1).

2. Before ordering the oligonucleotide primers, remember to include six additional nucleotides 5′ adjacent to the enzyme recognition site.
3. Have the oligonucleotide primers synthesized. An Applied Biosystems model 380A instrument was used for the synthesis of most of the oligonucleotide primers that we use. Purification of the oligonucleotide primers from residual organic molecules and pre-terminated (not full-length) oligonucleotide can be done by a number of standard methods that include polyacrylamide gel electrophoresis (PAGE) (DNA sequencing type gels, 7.5 M urea and 16% acrylamide), high performance liquid chromatography (HPLC), or the use of OPC purification cartridges (which can be obtained from Applied Biosystems, PN 400771).
4. The concentrations of the purified oligonucleotide primers are determined by measuring their optical density at 260 nm (for oligomers, an OD of 1 is equivalent with 20 μg/ml [30]).
5. Obtain a sample of the target DNA (plasmid, λ, genomic). The target DNA need not be extremely pure; for plasmid cloned DNAs we routinely use the crude plasmid DNA isolation procedure described by Morelle [32].
6. For each CPCR sample use a sterile 0.5 ml centrifuge tube and label it according to the CPCR sample being amplified. The order in which the reaction components are added is important (this is important for all PCR experiments [16, 21]); generally, the H_2O component is added first,

followed by the PCR buffer and the dNTP mix. The oligonucleotide primers and larget DNA are then added followed by the *Taq* polymerase.

7. Assuming the concentration of the target DNA is 100 μg/ml and that the concentration of each oligonucleotide primer has been adjusted to 100 μg/ml, the following amounts of each component are added. For the example presented here the CPCR sample contains the components for engineering the CaMV-CMV promoter and 5′-untranslated region of pUC1813cpCMV19.

Component	*Order of addition*	*Volume* (μl)
Double-distilled sterile H_2O	1	70.5
PCR Buffer (Perkin-Elmer-Cetus, or Solutions)	2	10.0
dNTP Mix, 1.25 mM (Perkin-Elmer-Cetus, or Solutions)	3	16.0
Primer JLS-81 (100 μg/ml)	4	1.0
Primer JLS-82 (100 μg/ml)	5	1.0
pUC1813cpCMV19 (100 μg/ml)	6	1.0
Taq polymerase (2.5 units/μl) (Perkin-Elmer-Cetus)	7	0.5
	Total	100.0

8. Mix gently, then briefly centrifuge in a microfuge. Overlay each PCR sample with 100 μl of mineral oil to prevent evaporation. The temperature control apparatus should be pre-set; if water baths are used they should be at the selected temperatures, and if an automated temperature cycler is used, the correct temperature control program should be selected.

   For the CPCR reactions used to engineer pUC18cpexp, a programmable cyclic reactor from Ericomp was used to control the following temperatures:

Duplex denaturation	94 °C	2 min
Annealing	50 °C	2 min
DNA synthesis	74 °C	3 min

9. Generally 20 to 30 cycles are used, and on the last cycle the denaturation step is omitted and the DNA synthesis step is extended to 10 min to allow

additional time to complete the synthesis of incomplete DNA strands. CPCR engineering of the DNA fragments used to construct the insert of pUC18cpexp involved the use of 30 cycles.

10. The yield of CPCR-amplified DNA products is analyzed by removing 10 μl from each sample and subjecting it to electrophoresis through a 7% acrylamide/15% glycerol gel. The amount of DNA product is estimated by staining the gel with ethidium bromide and observing the gel on an UV light box. If DNA bands, at the expected sizes, are observed and if their intensities suggest a sufficient yield (yields generally range between 0.2 to 2.0 μg) the CPCR products should be purified from the remaining sample. If the yield is low or no band is visible the experimental design should be rechecked (errors in designing the oligonucleotide primers are generally the problem). If the engineering designs are correct the CPCR amplification should be repeated.
11. If sufficient CPCR product is obtained, pipette off is much of the mineral oil as possible, then extract the remaining DNA product (90 μl) once with phenol chloroform isoamyl alcohol (1 : 1 : 0.04) and once with chloroform isoamyl alcohol (1 : 0.04). Add 9 μl of 3.0 M NaOAc (pH 6.3) and 300 μl of EtOH and chill at $-70\,^{\circ}$C for 10 min followed by pelleting the DNA product by certifuging it in an Eppendorf centrifuge at 4 °C. Wash once with 100% EtOH, and then dry under vacuum. Resuspend the CPCR DNA sample in 30 μl H_2O, and then digest the DNAs with the appropriate restriction enzyme(s).

    For this example both the 400 bp fragment containing the CaMV-CMV 35S promoter-5′-untranslated and 200 bp fragment containing the CaMV 35S terminator regions were first digested with *Nco* I. This would allow the construction of the 600 bp 35S CaMV-CMV cassette fragment by a single ligation event (Fig. 2).
12. After restriction enzyme digestion, remove all protein (restriction enzyme and BSA) by extracting once with phenol : chloroform : isoamyl alcohol, and once with chloroform : isoamyl alcohol. Then add 5 μl of sucrose loading dyes (which contains tracking dyes) and subject the DNA to electrophoresis through a 7% acrylamide/15% glycerol gel (if the CPCR fragments are larger than 1.5 kb they can be subjected to electrophoresis through a 0.7% agarose gel). CPCR DNA product bands are observed by staining with ethidium bromide.

    Fig. 2 (Frame A) shows the yields for the CPCR-amplified 400 bp CaMV-CMV promoter-5′-untranslated and 200 bp CaMV terminator fragments after electrophoresis through a 7% acrylamide/15% glycerol mini-gel (Hoefer Scientific Instrument, model SE200).

13. DNA bands of interest (CPCR products or phosphatase-treated vectors) are cut from the gel using either an unused razor blade or scalpel. DNAs are removed from the agarose or acrylamide gel matrix by electroelution as described by Maniatis *et al.* [30]. Tie one end of a narrow (0.5 cm wide) dialysis tube and place the cut gel part inside, then add about 200 μl of sterile 0.5× TBE electrophoresis buffer and tie the other end of the tube without trapping any air bubbles. Place the dialysis bag perpendicular to the electrodes of a horizontal agarose gel apparatus (such as a Hoefer Scientific Instrument, model HC-33) and submerge by adding sterile 0.5× TBE buffer. Electroelute the DNA band out of the gel matrix; an electric potential of about 100 V is sufficient to electroelute a 1 kb fragment in about 30 min. Movement of the DNA fragment out of the gel matrix can be observed by removing the top cover of the electrophoresis box (and thus electric potential) and exposing the dialysis bag to a hand-held UV light source. After the DNA has migrated out of the gel and appears to be pinned against the dialysis tubing, reverse the polarity of the electrodes and subject to electrophoresis for about 20 seconds to free the DNA from the surface of the dialysis tube. Remove the 200 μl of solution from the dialysis bag and transfer to a 1.5 ml centrifuge tube, then EtOH precipitate by adding 20 μl of 3 M NaOAc (pH 6.3), 600 μl of EtOH, chilling to −70 °C for 10 minutes, and precipitating the DNA pellet as described in step 11.
14. If the process requires the ligation of CPCR DNA products, estimate the amount of each DNA fragment recovered after electroelution by loading 1/10th of the sazmple onto an acrylamide glycerol gel (see step 10). The amount of DNA added to the ligation mix depends on the amount of DNA recovered (as estimated by the intensity of the ethidium bromide-stained band).

    For the construction of pUC18cpexp the 400 bp CaMV-CMV and 200 bp CaMV terminator fragments were ligated at the *Nco* I site. We estimated recovering about 1.0 μg of each CPCR DNA fragment, and they were resuspended in 30 μl of TEN buffer. The ligation reaction contained about 10 μl of each DNA fragment, which was transferred into a sterile 1.5 ml centrifuge tube followed by the addition of 2.3 μl of 10× ligation salts and 1 μl of ligase (10 units/μl; Collaborative Research, Inc.). This sample was gently mixed, briefly centrifuged and incubated at 14 °C for 18 h.

15. After ligation, remove the ligase by extracting once with phenol chloroform isoamyl alcohol, once with chloroform isoamyl alcohol, and EtOH-precipitate the DNA (step 13). Resuspend the vacuum-dried DNA pellet in 30 μl of sterile H_2O, and digest the CPCR-engineered DNA with the restriction enzyme whose site was synthesized into the flanking oligonucleotide primers. Generally use between 5 to 10 units of the enzyme per μg of DNA. After digestion, remove the restriction enzyme using the extraction procedure described in step 13, but do not EtOH-precipitate the DNA, instead add 3 μl of sucrose loading dye and electrophoretically separate the series of ligation products and non-ligated reagents on a 7% acrylamide/15% glycerol gel (step 12). Remove the appropriate ligation product from the gel matrix as described in step 13.

    The numerous ligation products resulting from the ligation of the two CPCR engineered DNA fragments for the construction of pUC18cpexp are shown in Fig. 2, Frame A. Many of the CPCR DNA fragments did not ligate, as indicated by the ethidium bromide staining intensities of the 200 bp CaMV terminator fragment (labeled T) and the 400 bp CaMV-CMV fragment (labeled P). A small amount of the ligated 600 bp CaMV-CMV promoter-terminator (PT) fragment is present. The other fragments present are the result of ligating two promoter fragments (800 bp) or two terminator fragments (most likely the lightly stained band located just above the 400 bp promoter fragment).

16. If the amount of the ligated products is sufficient (about 100 ng), proceed to step 17 for cloning into a plasmid vector. However, if the yield is very low or more product is needed for other experiments, it can be simply made by using a small amount (estimate 10 to 50 ng) of the ligated fragment as CPCR template. In this CPCR experiment add only the 5′- and 3′-flanking oligonucleotide primers. There is no need to repeat the original CPCR amplification processes to generate the individual DNA fragments. After a second round of CPCR amplification (steps 7 and 8), analyze and prepare the DNA fragment as described in steps 10 and 15, then proceed to step 17.

    For engineering the insert of pUC18cpexp, we did repeat the CPCR amplification to increase our yield of the 600 bp fragment. This was done by repeating steps 7 to 9, but only oligonucleotide primers JLS-81 and JLS-84 were added for CPCR amplification.

17. After obtaining sufficient quantities of the CPCR-engineered DNA fragment that has been digested with the appropriate restriction enzyme, clone it into the appropriate vector. If pUC18 is used, digest the plasmid

DNA with the appropriate restriction enzyme, adjust the pH to about 8.0 by the addition of 1/10th volume of 1 M Tris-HCl (pH 8.4), and remove the exposed 5′ phosphate groups by incubating the DNA at 55 °C for 30 min with 10 units per μg of DNA of calf intestinal phosphatase (Boehringer-Mannheim). Inactivate and remove proteins in the sample by extracting once with phenol chloroform isoamyl alcohol and once with chloroform isoamyl alcohol, and then subject the DNA to electrophoresis through a 7% acrylamide/15% glycerol gel (larger vectors should be purified using a 0.7% agarose gel). The linear plasmid DNA band is removed from the gel matrix using the electroelution procedure described in step 13.

18. Plasmid vector and insert DNA ligations are done by adding an estimated 50 ng of restriction enzyme cut-plasmid DNA and 50 ng of insert DNA, 1 μl of ligase salts into a sterile 1.5 ml centrifuge tube; the total volume is adjusted to 10 μl. The ligation reactions are incubated at 14 °C for 18 h. Following incubation, 3 μl of the ligation sample is added to 100 μl of competent *E. coli* DH5 or DH5α cells (Bethesda Research Laboratories), which are incubated at 37 °C for one hour and then plated on AB agar plates containing 100 μg/ml ampicillin.
19. Crude plasmid DNA preparations are made from selected colonies using the procedure described by Morelle [32]. These plasmid DNAs are first analyzed using the enzymes whose recognition sites were CPCR-engineered into the clone.

    For the identification of pUC18cpexp, plasmid DNAs were first digested with *Nco* I and *Hind* III. Five clones that showed the expected DNA fragment patterns (600 bp *Hind* III fragment and 400 and 200 bp *Nco* I + *Hind* III fragments) were completely sequenced; the sequence of the 600 bp insert from clone pUC18cpexp-17 is presented in Fig. 3.

Notes

2. The selection of the six nucleotides to add is not extremely important, but avoid the selection of homopolymers, palindromes, and the use of identical sequences for primers used in the same reaction. Once the oligonucleotide primers are designed, review their selection to ensure that the correct sequences are to be synthesized and that the expected DNA modifications are correct. Design errors are a waste of time and money (generally about $ 100 for each primer).
3. OPC cartridges are preferred because they are simple, rapid, and yield highly purified oligonucleotide primers. In addition, only full-length primers are purified because only trityl-bearing oligonucleotides are retained in the cartridge; pre-terminated oligonucleotides, which do not contain the trityl group, pass through the OPC cartridge. To avoid

the incorporation of a pre-terminated oligonucleotide in a CPCR product we recommend the use of OPC cartridges. However, for those who desire more adventure, the crude oligonucleotide primer mix can be used; oligonucleotides are supplied in concentrated NH_4OH, which is removed by vacuum drying followed by resuspending in TEN buffer.

5. In cases when the amount of target DNA is in question, a series of CPCR amplifications can be used to determine its effective concentration (the amount needed to yield a sufficient amount of CPCR product). We recommend that purified target DNAs be used to increase the chances of successfully obtaining a CPCR product.

10. The CPCR amplification should be repeated several times to ensure that the problem is not a simple pipeting error where one or more components were left out of the reaction. When repeating this CPCR application, use a series of different concentrations of the target DNA and/or individual oligonucleotide primers. If still no CPCR product can be obtained the oligonucleotide primers may be incorrect; they can either be sequenced using the chemical method [31] or replaced.

11. Restriction enzyme digestions are done using a 5- to 10-fold excess of enzyme using temperature and buffer conditions specified by the vendor.

17. This gel electrophoresis purification step removes traces of uncut plasmid and irreversibly denatured plasmid which is the result of using the alkaline plasmid purification method (plasmid DNA molecules that cannot be cut by restriction enzymes; Fig. 2, Frame B). This purification step greatly reduces the number of background colonies obtained when these vector DNAs are used to transform *E. coli* cells.

18, 19. Transformation of *E. coli* with the ligation mix described in step 18 yielded several hundred colonies. Forty of these colonies were selected (at random), 28 clones contained the correct 600 bp insert (for a dloning efficiency of 70%). Five of these clones were sequenced to check for ligation and/or PCR-generated nucleotide mutations. Crudely purified plasmid DNAs were sequenced using the double-stranded DNA sequence procedure described by Siemieniak *et al.* [48], and the 1 m gel system described by Slightom *et al.* [52], which is capable of obtaining average DNA readings of 800 bp. The complete 600 bp inserts were sequenced on both strands using the pUC18 universal primers. This analysis found no nucleotide sequence differences in the total 3000 bp that were sequenced. This result was encouraging because it indicates a very low rate of PCR-generated errors in these reactions. One reason for this low error rate is the use of large amounts of cloned template DNA, which effectively dilutes out any PCR-generated errors even if they occur early in the amplification process. Fig. 3 shows the locations of the CaMV 35S promoter, CMV 5′-untranslated region, *Nco* I cloning site, and CaMV poly(A) signal engineered into the constitutive expression vector referred to as pUC18cpexp. The use of CPCR engineering to modify the coding of a target gene for cloning into this vector is demonstrated in Section 2.

Solutions

- PCR dNTP mix (also obtained from Perkin-Elmer-Cetus)

– dATP 10mM	125 μl	
– dGTP 10mM	125 μl	
– dCTP 10mM	125 μl	
– dTTP 10mM	125 μl	
– H_2O	500 μl	Double-distilled sterile H_2O
Total	1.0 ml	dNTP mix, 1.25 mM final concentration for each dNTP

- PCR buffer 10× (also obtained from Perkin-Elmer-Cetus)

– KCl	500 mM
– Tris-HCl (pH 8.3)	100 mM
– $MgCl_2$	15 mM
– Gelatin	0.1% (w/v)

- Preparation of phenol extraction solution
  - Melt bottle of Mallinckrodt AR phenol at 68 °C.
  - Saturate with TEN buffer; about 250 ml/500 g phenol.
  - Add 2 M Tris-HCl (pH 9.5) until the pH of the phenol solution is between 7 and 8 (about 65 ml).
  - Add 0.2% 8-OH quinoline.
  - Add an equal volume of chloroform isoamyl alcohol (1 : 0.04) resulting in about 850 ml of phenol extraction solution.
- Glycerol acrylamide gel (7.5% acrylamide, 15% glycerol; 25 ml)
  - 7.5 ml of a 50% solution of glycerol (use 50% for ease of pouring)
  - 2.5 ml of a 10× Tris-borate-EDTA (TBE) electrophoresis buffer
  - 3.75 ml of a acrylamide stock solution (48% acrylamide, 2% bis)
  - 11.25 ml H_2O

  The gel is polymerized by the addition of 12 μl TEMED and 200 μl 10% ammonium persulfate.
- 10× TBE electrophoresis buffer (4 liters)

– Tris (4 M)	484.4 g
– Boric acid (3.3 M)	205.4 g crystals
– EDTA (5 mM)	14.9 g

  - Dissolve in a final volume of H_2O, check the pH, which should be about 8.3.

- Sucrose loading dye (100 ml)
  - sucrose 60% (w/v)
  - 10× TBE 5 ml
  - dyes 0.1 % xylene cyanol, bromphenol blue, orange G
- Preparation of CIP alkaline phosphatase
  - Obtain CIP from Boehringer Mannheim (No 405 612) (lyophilized) 4000 units
  - Resuspend in 1 ml glycerol buffer (10 mM Tris-HCl, pH 8.0; 1 mM $MgCl_2$; 40% glycerol), Carefully mix
  - Aliquot 100 μl into 1.5 ml Eppendorf tubes and store at −20 °C.
- TEN buffer
  - 10 mM Tris-HCl (pH 7.4)
  - 1 mM EDTA
  - 10 mM NaCl
- Ligase buffer 10×
  - 800 mM Tris-HCl (pH 8.0)
  - 200 mM $MgCl_2$
  - 150 mM DTT (note: DTT is saturated and at low temperature it will precipitate)
  - 10 mM ATP

```
       AAGCTTGCATGCCTGCAGGTCGACTCTAGAGGATCCCCGGGTACCGAGCTCGAATTGGGA
   1   ---------+---------+---------+---------+---------+---------+   60

       ATTCCCATGGAGTCAAAGATTCAAATAGAGGACCTAACAGAACTCGCCGTAAAGACTGGC
  61   ---------+---------+---------+---------+---------+---------+   120

       GAACAGTTCATACAGAGTCTCTTACGACTCAATGACAAGAAGAAAATCTTCGTCAACATG
 121   ---------+---------+---------+---------+---------+---------+   180

       GTGGAGCACGACACGCTTGTCTACTCCAAAAATATCAAAGATACAGTCTCAGAAGACCAA
 181   ---------+---------+---------+---------+---------+---------+   240

                                  HindIII
                                    ------
                    JLS-81   ACTAGGAAGCTT
                                         ---------------------
       AGGGCAATTGAGACTTTTCAACAAAGGGTAATATCCGGAAACCTCCTCGGATTCCATTGC
 241   ---------+---------+---------+---------+---------+---------+   300

       CCAGCTATCTGTCACTTTATTGTGAAGATAGTGGAAAAGGAAGGTGGCTCCTACAAATGC
 301   ---------+---------+---------+---------+---------+---------+   360

       CATCATTGCGATAAAGGAAAGGCCATCGTTGAAGATGCCTCTGCCGACAGTGGTCCCAAA
 361   ---------+---------+---------+---------+---------+---------+   420

       GATGGACCCCCACCCACGAGGAGCATCGTGGAAAAAGAAGACGTTCCAACCACGTCTTCA
 421   ---------+---------+---------+---------+---------+---------+   480

                           <----------------------CaMV Promoter-----
       AAGCAAGTGGATTGATGTGATATCTCCACTGACGTAAGGGATGACGCACAATCCCACTAT
 481   ---------+---------+---------+---------+---------+---------+   540

       ------------------------------------------------------------>
                             TATAA
       CCTTCGCAAGACCCTTCCTCTATATAAGGAAGTTCATTTCATTTGGAGAGGACAGGGTAC
 541   ---------+---------+---------+---------+---------+---------+   600

       <---------------------CMV 5'-Untranslated region-----------
       CCCCTACTGACTATATAGAGAGTGTGTGTGCTGTGTTTTCTCTTTTGTGTCGTAGAATTG
                                                       CAGCATCTTAAC
 601   ---------+---------+---------+---------+---------+---------+   660

       -------->

       AGTCGAGTCATGGACAAATCTGAATCAACCAGTGCTGGTCGTAACCATCGACGTCGTCCG
       TCAGCTCAG
               NcoI
              ------
            TTGGTACCGAACTG   JLS-82
 661   ---------+---------+---------+---------+---------+---------+   720
                MetAspLysSerGluSerThrSerAlaGlyArgAsnHisArgArgArgPro

       CGTCGTGGTTCCCGCTCCGCCCCCTCCTCCGCGGATGCTAACTTTAGAGTCTTGTCGCAG
 721   ---------+---------+---------+---------+---------+---------+   780
       ArgArgGlySerArgSerAlaProSerSerAlaAspAlaAsnPheArgValLeuSerGln

       CAGCTTTCGCGACTTAATAAGACGTTAGCAGCTGGTCGTCCAACTATTAACCACCCAACC
 781   ---------+---------+---------+---------+---------+---------+   840
       GlnLeuSerArgLeuAsnLysThrLeuAlaAlaGlyArgProThrIleAsnHisProThr

       TTTGTAGGGAGTGAACGCTGTAGACCTGGGTACACGTTCACATCTATTACCCTAAAGCCA
 841   ---------+---------+---------+---------+---------+---------+   900
       PheValGlySerGluArgCysArgProGlyTyrThrPheThrSerIleThrLeuLysPro

       CCAAAAATAGACCGTGAGTCTTATTACGGTAAAAGGTTGTTACTACCTGATTCAGTCACG
 901   ---------+---------+---------+---------+---------+---------+   960
       ProLysIleAspArgGluSerTyrTyrGlyLysArgLeuLeuLeuProAspSerValThr

       GAATATGATAAGAAGCTTGTTTCGCGCATTCAAATTCGAGTTAATCCTTTGCCGAAATTT
 961   ---------+---------+---------+---------+---------+---------+   1020
       GluTyrAspLysLysLeuValSerArgIleGlnIleArgValAsnProLeuProLysPhe
```

Fig. 1. For caption see overleaf.

```
        GATTCTACCGTGTGGGTGACAGTCCGTAAAGTTCCTGCCTCCTCGGACTTATCCGTTGCC
  1021  ---------+---------+---------+---------+---------+---------+   1080
        AspSerThrValTrpValThrValArgLysValProAlaSerSerAspLeuSerValAla

        GCCATCTCTGCTATGTTCGCGGACGGAGCCTCACCGGTACTGGTTTATCAGTATGCCGCA
  1081  ---------+---------+---------+---------+---------+---------+   1140
        AlaIleSerAlaMetPheAlaAspGlyAlaSerProValLeuValTyrGlnTyrAlaAla

        TCTGGAGTCCAAGCCAACAACAAACTGTTGTTTGATCTTTCGGCGATGCGCGCTGATATA
  1141  ---------+---------+---------+---------+---------+---------+   1200
        SerGlyValGlnAlaAsnAsnLysLeuLeuPheAspLeuSerAlaMetArgAlaAspIle

        GGTGACATGAGAAAGTACGCCGTCCTCGTGTATTCAAAAGACGATGCGCTCGAGACGGAC
  1201  ---------+---------+---------+---------+---------+---------+   1260
        GlyAspMetArgLysTyrAlaValLeuValTyrSerLysAspAspAlaLeuGluThrAsp

        GAGCTAGTACTTCATGTTGACATCGAGCACCAACGCATTCCCACATCTGGAGTGCTCCCA
  1261  ---------+---------+---------+---------+---------+---------+   1320
        GluLeuValLeuHisValAspIleGluHisGlnArgIleProThrSerGlyValLeuPro

        GTCTGATTCCGTGTTCCCAGAACCCTCCCTCCGATCTCTGTGGCGGGAGCTGAGTTGGCA
  1321  ---------+---------+---------+---------+---------+---------+   1380
        ValEnd

        GTTCTACTACAAACTGTCTGGAGTCACTAAACGTTTTACGGTGAACGGGTTGTCCATCCA
  1381  ---------+---------+---------+---------+---------+---------+   1440

        GCTTACGGCTAAAATGGTCAGTCGTGGAGAAATCCACGCCAGCAGATTTACAAATCTCTG
  1441  ---------+---------+---------+---------+---------+---------+   1500

        AGGCGCCTTTGAAACCATCTCCTAGGTTTCTTCGGAAGGGCTTCGGTCCGTGTACCTCTA
  1501  ---------+---------+---------+---------+---------+---------+   1560

        GCGCAACGTGCTAGTTTCAGGGTACGGGTGCCCCCCCACTTTCGTGGGGGCCTCCAAAAG
  1561  ---------+---------+---------+---------+---------+---------+   1620

        GAGACCAAAAAAAAAAAAAAAAAAAAAAAAAAAAAAAAAAAAAAAAAAAAAAAAAAAAAA
  1621  ---------+---------+---------+---------+---------+---------+   1680

        AAAAAAAAAAAAAAAAAAAAAAAAAAAAAAAAAGAATTGGGGATCCTCTAGAGTCGACCTG
  1681  ---------+---------+---------+---------+---------+---------+   1740
GCAATTCCATGGTT
      ------
       NcoI
               ----------------------  JLS-83
        CAGGCATGCGCTGAAATCACCAGTCTCTCTCTACAAATCTATCTCTCTCTATAATAATGT
  1741  ---------+---------+---------+---------+---------+---------+   1800

        GTGAGTAGTTCCCAGATAAGGGAATTAGGGTTCTTATAGGGTTTCGCTCATGTGTTGAGC
  1801  ---------+---------+---------+---------+---------+---------+   1860

                                                   CaMV Poly (A) signal
                                                         ------
        ATATAAGAAACCCTTAGTATGTATTTGTATTTGTAAAATACTTCTATCAATAAAATTTCT
  1861  ---------+---------+---------+---------+---------+---------+   1920

        AATTCCTAAAACCAAAATCCAGTGGTACCGAGCTCGAATTCGAGCTCCCCGGGGATCCTC
                 TTGGTTTTAGGTCACCATGG
                                     HindIII
                                     ------
                                     TTCGAATCAAGC  JLS-84
  1921  ---------+---------+---------+---------+---------+---------+   1980

        TAGAGTCGACCTGCAGGCATGCAAGCTT
  1981  ---------+---------+--------  2008
```

Fig. 1. Nucleotide sequence of the insert of clone pUC1813cpCMV19, which was engineered for the expression of the CMV-C coat protein gene [41]. Nucleotide sequences important for expression, the CaMV promoter and terminator, and the CMV 5′-untranslated regions are indicated. The location and sequence of the four oligonucleotide primers used for PCR-engineering the expression

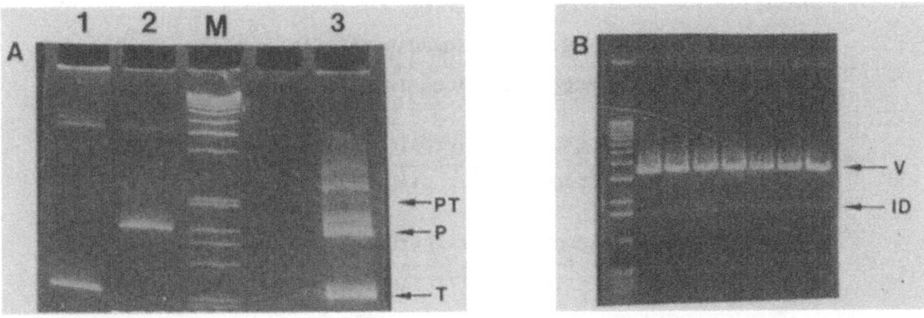

Fig. 2. CPCR amplification of promoter and terminator regions for the construction of pUC18cpexp. Frame A shows the amplification of CaMV terminator (lane 1) and CaMV-CMV promoter (lane 2) regions. The lane labeled M contains the 1 kb DNA size standard from Bethesda Research Laboratories. The sequence and position of the oligonucleotide primers used for this amplification are shown in Fig. 1. Lane 3 shows the results of ligating the 200 bp and 400 bp fragments (after digestion with *Nco* I); many of the fragments did not ligate as indicated by the presence of the 200 bp terminator fragment (labeled T) and 400 bp promoter fragment (labeled P). Ligation products show the presence of the 600 bp promoter and terminator fragment (labeled PT) and other combinations, 400 bp fragment (ligation of two terminator fragments) and 800 bp fragment (ligation of two promoter fragments). Frame B shows the vector pUC18 after digestion with *Hind* III (see Section 1, step 17). The vector DNA band is indicated by the letter V and the less intense band, which is irreversibly denatured plasmid DNA, is indicated by the letters ID. Removal of the irreversibly denatured DNA is important to reduce the bacterial colony background during *E. coli* transformation experiments (see Section 1, steps 17 and 18).

---

cassette pUC18cpexp (see Fig. 3) are shown along with the restriction enzyme recognition site sequences (internal *Nco* I and flanking *Hind* III sites) synthesized within each oligonucleotide primer. Dashed lined regions of the oligonucleotide primers JLS-81 and -83 indicate shared nucleotide identity with the template. The complementary sequences are shown for the oligonucleotide primers JLS-82 and -84.

```
       HindIII
       -------
       GAAGCTTCCGGAAACCTCCTCGGATTCCATTGCCCAGCTATCTGTCACTTTATTGTGAAG
   1   ---------+---------+---------+---------+---------+---------+   60
       CTTCGAAGGCCTTTGGAGGAGCCTAAGGTAACGGGTCGATAGACAGTGAAATAACACTTC

       ATAGTGGAAAAGGAAGGTGGCTCCTACAAATGCCATCATTGCGATAAAGGAAAGGCCATC
  61   ---------+---------+---------+---------+---------+---------+   120
       TATCACCTTTTCCTTCCACCGAGGATGTTTACGGTAGTAACGCTATTTCCTTTCCGGTAG

       GTTGAAGATGCCTCTGCCGACAGTGGTCCCAAAGATGGACCCCCACCCACGAGGAGCATC
 121   ---------+---------+---------+---------+---------+---------+   180
       CAACTTCTACGGAGACGGCTGTCACCAGGGTTTCTACCTGGGGGTGGGTGCTCCTCGTAG

                                                             <------
       GTGGAAAAAGAAGACGTTCCAACCACGTCTTCAAAGCAAGTGGATTGATGTGATATCTCC
 181   ---------+---------+---------+---------+---------+---------+   240
       CACCTTTTTCTTCTGCAAGGTTGGTGCAGAAGTTTCGTTCACCTAACTACACTATAGAGG

       --------------------CaMV Promoter---------------------------
       ACTGACGTAAGGGATGACGCACAATCCCACTATCCTTCGCAAGACCCTTCCTCTATATAA
 241   ---------+---------+---------+---------+---------+---------+   300
       TGACTGCATTCCCTACTGCGTGTTAGGGTGATAGGAAGCGTTCTGGGAAGGAGATATATT

       --------------------------------->
                                          <-------------------------
       GGAAGTTCATTTCATTTGGAGAGGACAGGGTACCCCTACTGACTATATAGAGAGTGTGTG
 301   ---------+---------+---------+---------+---------+---------+   360
       CCTTCAAGTAAAGTAAACCTCTCCTGTCCCATGGGGATGACTGATATATCTCTCACACAC

                                                   NcoI
                                                  ------
       ----CMV 5'-Untranslated Region--------------->
                                                    ATG
       TGCTGTGTTTTCTCTTTTGTGTCGTAGAATTGAGTCGAGTCAACCATGGTTGCGCTGAAA
 361   ---------+---------+---------+---------+---------+---------+   420
       ACGACACAAAAGAGAAAACACAGCATCTTAACTCAGCTCAGTTGGTACCAACGCGACTTT

       TCACCAGTCTCTCTCTACAAATCTATCTCTCTCTATAATAATGTGTGAGTAGTTCCCAGA
 421   ---------+---------+---------+---------+---------+---------+   480
       AGTGGTCAGAGAGAGATGTTTAGATAGAGAGAGATATTATTACACACTCATCAAGGGTCT

       TAAGGGAATTAGGGTTCTTATAGGGTTTCGCTCATGTGTTGAGCATATAAGAAACCCTTA
 481   ---------+---------+---------+---------+---------+---------+   540
       ATTCCCTTAATCCCAAGAATATCCCAAAGCGAGTACACAACTCGTATATTCTTTGGGAAT

                                  CaMV Poly (A) signal
                                        ------
       GTATGTATTTGTATTTGTAAAATACTTCTATCAATAAAATTTCTAATTCCTAAAACCAAA
 541   ---------+---------+---------+---------+---------+---------+   600
       CATACATAAACATAAACATTTTATGAAGATAGTTATTTTAAAGATTAAGGATTTTGGTTT

                   HindIII
                   -------
       ATCCAGTGGTACCAAGCTTC
 601   ---------+---------+   620
       TAGGTCACCATGGTTCGAAG
```

Fig. 3. Nucleotide sequence of the CPCR-engineered expression cassette pUC18cpexp-17, CPCR-engineered in Section 1. The location of important CaMV-CMV gene regulatory elements and restriction enzyme sites are indicated. The oligonucleotide primers used for CPCR amplification are presented in Fig. 1.

Section 2. CPCR engineering of plant virus coat protein genes

The expression cassette, pUC18cpexp, can be used to promote constitutive expression of gene coding regions that do not contain a 5′-untranslated region or even a translation initiation site. In this example, we present the use of CPCR engineering to modify the gene encoding the coat protein of a plant virus belonging to the potyvirus group, watermelon mosaic virus II (WMVII). Potyvirus proteins are encoded as a polyprotein [13, 55]; their individual proteins are cleaved from the polyprotein by proteases that are also encoded in the polyprotein [8, 9, 12]. The potyvirus coat protein resides as the 3′-most protein of the polyprotein [1, 12, 29, 39, 40, 43], thus it does not contain a 5′-untranslated region or even a translation initiation site.

The expression cassette, pUC18cpexp, engineered in Section 1 is well suited for expressing potyviral coat protein genes, because once they are cloned into pUC18cpexp (after the addition of flanking *Nco*I sites by CPCR engineering) they are provided with the necessary translational regulatory elements. The design of the oligonucleotide primers needed to add *Nco*I sites flanking the WMVII coat protein gene is for the most part straightforward, except that some thought must be put into the selection of which WMVII coat protein codon should be fused to the CMV translation initiation codon (ATG). The criteria and thought processes used for this selection are described in this section.

2.1. CPCR engineering of the WMVII coat protein gene and cloning into pUC18cpexp

Steps in the procedure

1. The goal of this example is to obtain a translational fusion between pUC18cpexp and the gene encoding the WMVII coat protein that will produce a protein as nearly identical to the native WMVII coat protein as possible. The complete amino acid and nucleotide sequences of the WMVII coat protein have been determined [40, 55]; its N-terminal amino acid, Gly, is proceeded by a Ser residue, not a Met. Thus some modification of the WMVII coat protein is unavoidable; in addition, the fusion cannot be made to the native coat protein N-terminus. A reduction in the length of the coat protein is needed to preserve the consensus translation initiation signal (AAXXATGG, contained in the 5′ oligonucleotide primer); it is more convenient to fuse it to a codon beginning with a G nucleotide. In addition, because the N-terminus of most potyvirus coat proteins is either a Ser, Gly, or Ala residue [14], these residues are preferentially selected. The addition

of the *Nco* I site to the 3′ end of these coat protein gene is much simpler because the linkage involves a transcriptional fusion.

2. Oligonucleotides designed for CPCR amplification of the WMVII coat protein gene fuse the translation initiation codon to the second amino acid codon, Gly, and add the 3′ *Nco* I site 40 nucleotides downstream of the WMVII coat protein translation termination codon. The oligonucleotide primers used for CPCR engineering the WMVII coat protein gene are shown in Fig. 4. The WMVII coat protein gene clone pWMVII-3.2 [40] supplied the necessary template, and the CPCR amplification was done as described in Section 1 (steps 7 and 8).
3. CPCR amplification yielded a 900 bp fragment that was digested with *Nco* I and then purified by electrophoresis through an acrylamide gel followed by electroelution from the gel matrix (Section 1, steps 10 to 13). The CPCR DNA product was cloned into *Nco* I-cut pUC18cpexp (the vector was prepared as described in Section 1, step 17) as described in Section 1, step 18. The complete ligation mix was used to transform *E. coli* cells (three separate transformation experiments) as described in Section 1 (step 18) and only five clones were obtained. These clones were mapped with *Hin*d III, *Nco* I, and/or *Xho* I. *Xho* I was used to determine the orientation of the cloned WMVII insert; only one clone contained the WMVII coat protein gene insert in the 'N' orientation (see Notes); this clone is referred to as pUC18cpWMVII-5.
4. Nucleotide sequence analysis of this insert showed that it was correctly fused to the translation initiation codon and that it contained no PCR-generated or ligation errors; the nucleotide sequence of the insert of pUC18cpWMVII-5 is shown in Fig. 4.

Notes

3. Restriction enzyme site mapping of inserts cloned into pUC18cpexp is necessary to determine the orientation of the insert with respect to the CaMV 35S promoter. The orientation of inserts can be determined using a restriction enzyme site that is located within the insert and mapping it with respect to a site in the pUC18 polylinker. Cloned inserts found in the same 5′ to 3′ orientation as the CaMV-CMV 35S regulatory sequences are referred to as being cloned in the 'N' orientation and those in the opposite direction are referred to being cloned in the 'U' orientation. Discard the clones in the U orientation unless they are the only type that are obtained; they can be digested and used as a cloned source of the insert to obtain the desired N-oriented clone.
4. We have had considerable success in using the CPCR engineering technique to clone potyviral coat protein genes into pUC18cpexp; we have cloned coat protein genes from papaya ringspot virus (PRV) and zucchini yellow mosaic virus (ZYMV) into this vector. The engineering process for the PRV coat protein gene [53] fused the translation initiation

codon to the fifth amino acid (Ala) that follows the expected Glu-Ser peptide cleavage site [39]. The 3′ *Nco* I site was added 50 bp downstream of the PRV translation terminator codon. Transformation of *E. coli* (after ligating the CPCR engineered fragment into pUC18cpexp) yielded 117 colonies, of which 22 out of 24 contained the expected size insert, and eight of these were in the correct N orientation. The coat protein gene of ZYMV was CPCR-engineered [53] by fusing the translation initiation codon to the second amino acid codon (Gly) following the expected Glu-Ser peptide cleavage site [40]. The 3′ *Nco* I site was added 60 bp downstream from the ZYMV translation termination codon. About 150 independent clones were obtained, and 16 out of 18 checked (by digestion with *Hind* III and *Nco* I) contained the expected insert fragment. Five out of the 16 clones contained the insert in the N orientation. The nucleotide sequence of the first two selected clones did contain cloning artifacts that appeared to be the result of ligation errors rather than PCR-generated errors. Analysis of an additional clone yielded the expected CPCR-engineered and ligation insert. The presence of these artifacts clearly demonstrate the need for detailed analysis of any engineered clone prior to embarking on the difficult and time consuming task of transferring it into a plant species and later finding no expression.

Our ability rapidly to engineer three coat protein genes (all of which originally did not contain 5′-untranslated regions and translation initiation codons) for expression in plants demonstrates the power of CPCR engineering. Although the components of CPCR engineering are not completely error-free, its ability to reduce complex DNA engineering to a series of simple and straightforward processes, such as DNA ligation, clearly offers the molecular biologists advantages that outweigh its inherent weaknesses.

Transfer of these plant virus coat protein gene expression cassettes into other plasmid constructions can be accomplished using the flanking CPCR-engineered *Hind* III sites (see Figs. 3 and 4). The *Hind* III site was selected because it is not present in the PRV, WMVII, or ZYMV coat protein genes, and because the *Hind* III site is present in the widely used *Agrobacterium* binary vector pGA482 polylinker [4]. Changes in this cloning strategy can quickly be achieved by cloning the expression cassettes into a plasmid that contains different polycloning sites, such as pUC1813 or pUC1318 [24] or by using CPCR-engineering to add different restriction enzyme sites flanking the expression cassette. Transfer of these plant virus expression cassettes into the modified *Agrobacterium* binary vector pGA482GG [41] has been accomplished, and these vectors were subsequently transferred into avirulent *Agrobacterium* strains C58Z707 [20]. Each of these CPCR-engineered potyvirus coat protein expression cassettes has been tested and found to produce high levels of their respective potyvirus coat protein [35].

```
       HindIII
       ------
       GAAGCTTCCGGAAACCTCCTCGGATTCCATTGCCCAGCTATCTGTCACTTTATTGTGAAG
   1   ---------+---------+---------+---------+---------+---------+   60

       ATAGTGGAAAAGGAAGGTGGCTCCTACAAATGCCATCATTGCGATAAAGGAAAGGCCATC
  61   ---------+---------+---------+---------+---------+---------+  120

       GTTGAAGATGCCTCTGCCGACAGTGGTCCCAAAGATGGACCCCCACCCACGAGGAGCATC
 121   ---------+---------+---------+---------+---------+---------+  180
                                                             <------
       GTGGAAAAAGAAGACGTTCCAACCACGTCTTCAAAGCAAGTGGATTGATGTGATATCTCC
 181   ---------+---------+---------+---------+---------+---------+  240

       ------------CaMV Promoter-----------------------------------
       ACTGACGTAAGGGATGACGCACAATCCCACTATCCTTCGCAAGACCCTTCCTCTATATAA
 241   ---------+---------+---------+---------+---------+---------+  300

       ----------------------------->
                                      <-----------------------
       GGAAGTTCATTTCATTTGGAGAGGACAGGGTACCCCTACTGACTATATAGAGAGTGTGTG
 301   ---------+---------+---------+---------+---------+---------+  360

       ----CMV 5'-Untranslated Region---------------->
                                                 NcoI
                                                 ------
                                    JLS-87   ATCATTCCATGG
                                                   ------------
       TGCTGTGTTTTCTCTTTTGTGTCGTAGAATTGAGTCGAGTCAACCATGGGAAAAGAAACA
 361   ---------+---------+---------+---------+---------+---------+  420
                                                  MetGlyLysGluThr

       --------
       GTTGAAAATTTGGACGCAGGGAAAGAATCAAAGAAAGATGCCAGTGACAAAGGGAATAAG
 421   ---------+---------+---------+---------+---------+---------+  480
       ValGluAsnLeuAspAlaGlyLysGluSerLysLysAspAlaSerAspLysGlyAsnLys

       CCGCAGAACTCGCAAGTTGGTCAGGGTAGCAAGGAACCAACAAAAACCGGCACAGTCAGC
 481   ---------+---------+---------+---------+---------+---------+  540
       ProGlnAsnSerGlnValGlyGlnGlySerLysGluProThrLysThrGlyThrValSer

       AAGGATGTAAATGTTGGATCGAAAGGAAAAGAAGTCCCACGACTACAAAAGATAACAAAG
 541   ---------+---------+---------+---------+---------+---------+  600
       LysAspValAsnValGlySerLysGlyLysGluValProArgLeuGlnLysIleThrLys

       AAAATGAATCTTCCAACAGTTGGTGGAAAATCATTCTTAGCTTAGACCATTTGCTTGAG
 601   ---------+---------+---------+---------+---------+---------+  660
       LysMetAsnLeuProThrValGlyGlyLysIleIleLeuSerLeuAspHisLeuLeuGlu

       TACAAACCTAGTCAAGTTGATTTGTTTAACACTCGAGCAACAAAAACACAATTTGAATCA
 661   ---------+---------+---------+---------+---------+---------+  720
       TyrLysProSerGlnValAspLeuPheAsnThrArgAlaThrLysThrGlnPheGluSer

       TGGTACAGCGCAGTCAAAGTTGAGTATGATCTTAATGATGAGCAAATGGGTGTGATTATG
 721   ---------+---------+---------+---------+---------+---------+  780
       TrpTyrSerAlaValLysValGluTyrAspLeuAsnAspGluGlnMetGlyValIleMet

       AATGGTTTTATGGTTTGGTGTATCGATAACGGTACATCTCCAGATGTCAATGGAGTGTGG
 781   ---------+---------+---------+---------+---------+---------+  840
       AsnGlyPheMetValTrpCysIleAspAsnGlyThrSerProAspValAsnGlyValTrp

       GTAATGATGGATGGGGAAGAGCAAGTTGAGTACCCACTAAAGCCAATTGTTGAAAATGCA
 841   ---------+---------+---------+---------+---------+---------+  900
       ValMetMetAspGlyGluGluGlnValGluTyrProLeuLysProIleValGluAsnAla

       AAGCCAACTTTGAGACAAATCATGCACCATTTCTCAGATGCAGCGGAAGCATATATTGAA
 901   ---------+---------+---------+---------+---------+---------+  960
       LysProThrLeuArgGlnIleMetHisHisPheSerAspAlaAlaGluAlaTyrIleGlu

       ATGAGAAACTCTGAAAGTCCGTATATGCCTAGATACGGATTACTGAGAAATTTGAGAGAC
 961   ---------+---------+---------+---------+---------+---------+ 1020
       MetArgAsnSerGluSerProTyrMetProArgTyrGlyLeuLeuArgAsnLeuArgAsp
```

Fig. 4.

```
      AGGGAATTAGCACGCTATGCTTTTGACTTCTATGAGGTTACTTCTAAAACGCCAAATAGG
1021  ---------+---------+---------+---------+---------+---------+  1080
      ArgGluLeuAlaArgTyrAlaPheAspPheTyrGluValThrSerLysThrProAsnArg

      GCAAGAGAAGCAATAGCTCAAATGAAGGCCGCAGCTCTCGCGGGAGTTAACAGCAGGTTA
1081  ---------+---------+---------+---------+---------+---------+  1140
      AlaArgGluAlaIleAlaGlnMetLysAlaAlaAlaLeuAlaGlyValAsnSerArgLeu

      TTTGGACTTGATGGTAATATCTCGACCAATTCCGAAAATACTGGGAGGCACACTGCAAGG
1141  ---------+---------+---------+---------+---------+---------+  1200
      PheGlyLeuAspGlyAsnIleSerThrAsnSerGluAsnThrGlyArgHisThrAlaArg

      GACGTAAATCAGAATATGCATACTTTGTTGGGTATGGGTCCACCGCAGTAAAGACTAGGT
1201  ---------+---------+---------+---------+---------+---------+  1260
      AspValAsnGlnAsnMetHisThrLeuLeuGlyMetGlyProProGlnEnd

                                    NcoI
                                    ------
      AAACTGGTCACAGTTAGCATTTCGGGTCGCCATGGTTGCGCTGAAATCACCAGTCTCTCT
              GTGTCAATCGTAAAGCCCAGCGGTACC
                                         AATCGA  JLS-88
1261  ---------+---------+---------+---------+---------+---------+  1320

      CTACAAATCTATCTCTCTCTATAATAATGTGTGAGTAGTTCCCAGATAAGGGAATTAGGG
1321  ---------+---------+---------+---------+---------+---------+  1380

      TTCTTATAGGGTTTCGCTCATGTGTTGAGCATATAAGAAACCCTTAGTATGTATTTGTAT
1381  ---------+---------+---------+---------+---------+---------+  1440

                CaMV Poly (A) signal
                        ------                                   -
      TTGTAAAATACTTCTATCAATAAAATTTCTAATTCCTAAAACCAAAATCCAGTGGTACCA
1441  ---------+---------+---------+---------+---------+---------+  1500

      HindIII
      -----
      AGCTTC
1501  ------  1506
```

Fig. 4. Nucleotide sequence of CPCR-engineered expression vector pUC18cpWMVII-5. Nucleotide sequence elements important for the expression of the WMVII coat protein gene are indicated, along with the sequence and location of oligonucleotide primers used for CPCR-amplification. The location of the *Nco* I sites used for cloning the CPCR amplified WMVII coat protein gene into pUC18cpexp are shown. Shared nucleotide sequences between the oligonucleotide primers and WMVII coat protein gene are shown as dashed lines above (primer JLS-87) and as complementary DNA strand below (JLS-88) the presented nucleotide sequence strand.

Section 3. CPCR engineering of a seed-specific expression cassette from cloned DNA

CPCR engineering can be used to construct expression cassettes using the gene regulatory sequences of almost any gene. This technology has few limitations except size (DNA fragments larger than 5 kb produce low CPCR yields) and restriction enzyme site content, which could make the selection of CPCR-added restriction enzyme sites difficult (see Sections 1 and 2). In the following example we demonstrate the use of CPCR engineering to construct an expression vector using the regulatory regions of the *Phaseolus vulgaris* seed storage protein gene, phaseolin. Slightom *et al.* [50] reported the complete nucleotide sequence of the phaseolin gene, including 5′- and 3′-regulatory sequences and its cDNA counterpart, and Doyle *et al.* [15] reported an additional 800 bp of the phaseolin 5′-flanking DNA. The promoter region of the phaseolin gene is similar to that identified for many other plant genes (containing CCAAT and TATAA elements in its 5′-flanking region and the AATAAA polyadenylation signal in its 3′-untranslated region). The sequence and location of other potential regulatory elements that may be responsible for developmental regulation of the phaseolin gene have also been identified [11]. The phaseolin-coding region encodes a signal peptide of about 22 to 25 amino acid residues [50, 51], which complicates the decision on how the expression cassette should be engineered; should the signal peptide be included? It would not be surprising to find cases where the signal peptide would be useful and others where it would not. A similar decision must be made in engineering the translation and transcriptional portion of this cassette; that is, will a 3′ translation fusion be needed (which would include the use of the phaseolin translational termination codon), or is only the polyadenylation signal AATAAA needed? There is simply a need for two types of a seed-specific expression cassette, one for expressing coding regions that require a signal peptide and translational termination codon (this is referred to as the 'pvPSPTexp' insert) and another cassette for the expression of genes that only need a promoter and polyadenylation signal (this cassette is referred to as the 'pvPTexp' insert). The cassette, pvPSPTexp, will also be useful for the expression of coding regions whose native signal peptide must be removed because, if left intact it may result in incorrect targeting of the synthesized polypeptide.

3.1. CPCR engineering seed expression vectors based on the phaseolin regulatory sequences

Steps in the procedure

1. From the nucleotide sequence of the phaseolin gene [15, 50] we determined strategies to CPCR-engineer both phaseolin expression cassettes. The expression cassettes will be flanked by *Hind* III sites and a single internal *Nco* I site will be used for cloning target coding regions. The phaseolin sequence does not contain any *Hind* III sites and it contains only one *Nco* I site, located 500 bp 5′ of the translation initiation site.
2. CPCR engineering of both phaseolin expression cassettes will require the synthesis of six oligonucleotide primers, and the position of these oligonucleotides is shown in Fig. 5. The oligonucleotide primers listed below were selected for the following reasons:
   – Primer JLS-93 (5′-CCTAAGAAGCTTGGCACTCTGTGGTCTTTTGGTTC-3′) places a *Hind* III site at the location formerly occupied by the *Nco* I site (500 bp 5′ of the phaseolin translation initiation signal), thus preventing the incorporation of an unwanted *Nco* I site. This primer can be used to CPCR-engineer both phaseolin expression cassettes. Each cassette will contain 424 bp of 5′-flanking DNA.
   – Primer JLS-94 (5′-GTCAAGCCATGGTTAGTAGAGTAGTATTGAATA-TG-3′) places a *Nco* I site at the phaseolin translation codon. The design of this oligonucleotide leaves only one Met residue at the phaseolin translation initiation codon, thus eliminating the possibility of having two potential translation initiation codons [50, 51]. This CPCR engineered translation initiation codon also conforms to the consensus sequence [26, 27].
   – Primer JLS-95 (5′-CAATTGCCATGGAAGTATGAACTAAAATGCATG-TAG-3′) places an *Nco* I site 4 bp 5′ of the phaseolin terminator codon.
   – Primer JLS-96 (5′-CAACTGAAGCTTGTTGAAACATCCCTGAAGTGTC-3′) places a *Hind* III site 1170 bp 3′ of the phaseolin termination codon.
   – Primer JLS-133 (5′-AATTAATCCATGGCCGGAGTGAAGTGGCAAA-TG-3′) places a *Nco* I site 12 bp (4 amino acids) 5′ of the last amino acid cleavage point for the phaseolin signal peptide [50, 51].
   – Primer JLS-134 (5′-AATCTTCCATGGGCATTTGTGTACTGAATAAGT-ATG-3′) places a *Nco* I site 12 bp (4 amino acids) 3′ of the phaseolin translation termination codon.
3. For each CPCR amplification sample, use a sterile 0.5 ml centrifuge tube labeled according to the CPCR sample number. In this example four CPCR amplifications are needed to obtain both CPCR-engineered phaseolin cas-

settes. The phaseolin clone pPvPh3.8β[10] was used to supply the phaseolin gene regulatory sequence template for each of the CPCR amplifications. About 100 ng of pPvPh3.8β DNA and 50 ng of each oligonucleotide primer was used, and the CPCR amplifications were done using the PCR routine described in Section 1 (steps 7 to 9). The following CPCR amplifications were performed:

– oligonucleotide primers JLS-93 and -94 produce a 508 bp fragment containing the phaseolin promoter (see Fig. 6);

– oligonucleotide primers JLS-95 and -96 produce a 1170 bp fragment containing the phaseolin polyadenylation signal (see Fig. 6);

– oligonucleotide primers JLS-93 and -133 produce a 590 bp fragment containing the phaseolin promoter and signal peptide (see Fig. 6);

– oligonucleotide primers JLS-134 and -96 yield a 1195 bp fragment containing the phaseolin terminator and polyadenylation signal (see Fig. 6).

4. The CPCR samples were checked for product yield as described in Section 1 (step 10). *Taq* polymerase was removed and the DNA digested with the appropriate restriction enzyme (Section 1, step 11); in this case, the *Nco* I sites were exposed by digestion, and these *Nco* I-cut DNAs were purified by electrophoresis through a 7% acrylamide/15% glycerol gel shown in Fig. 6 (see Section 1, steps 12 and 13). The phaseolin expression cassettes were assembled by ligating the appropriate CPCR-engineered fragments (see Section 1, step 14) as follows. The CPCR-engineered fragments a and b (step 3) were ligated to obtain the 1670 bp phaseolin cassette fragment pvPexp and CPCR-engineered fragments c and d were ligated to obtain the 1785 bp phaseolin cassette fragment pvPSPTexp.
5. The pvPTexp ligation product was treated as described in Section 1, step 15, to remove the ligase, to expose the flanking CPCR-added *Hind* III sites, and to purify the correct size CPCR-engineered phaseolin cassette fragments from non-ligated and other ligation products. The phaseolin cassette fragment was removed from the acrylamide gel matrix as described in Section 1, step 13. However, the pvPSPTexp ligation product yield was very low. It was isolated and used as template for a second round of CPCR amplification using the flanking oligonucleotide primers JLS-93 and -96, as described in Section 1, step 16. The products of this second round of CPCR amplification are shown in Fig. 7.
6. The purified phaseolin cassette fragments were cloned into the same *Hind* III-digested pUC18 vector described in Section 1 (steps 18), using the ligation reaction described in Section 1 (step 18). Transformation of competent *E. coli* cells with the ligation products pvPTexp and pvPSPT exp

yielded 32 and several hundred independent clones, respectively. Eighteen colonies were selected from transformants containing the pvPTexp cassette and 36 from transformants containing the pvPSPTexp cassette and they were subjected to restriction enzyme mapping. Those that showed the correct size cloned fragments were further analyzed by obtaining their nucleotide sequences. Clones referred to as pUC18pvPTexp7-7, 7-10, and 7-14 and pUC18pvPSPTexpD22 and D23 were found to contain the expected CPCR-engineered phaseolin gene regulatory regions. The nucleotide sequence of pUC18pvPSPTexpD22 is presented in Fig. 8.

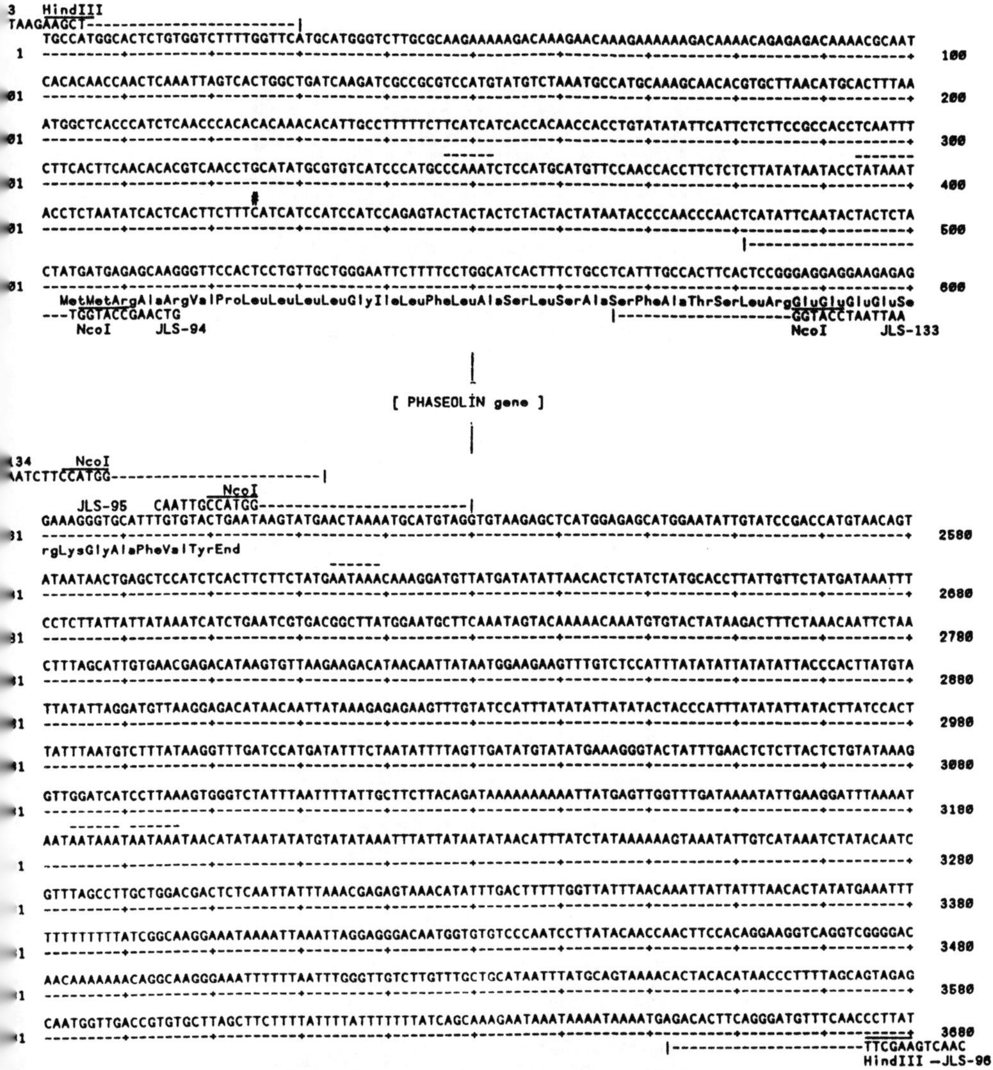

Fig. 5. CPCR engineering of the phaseolin seed storage protein expression cassettes inserts, pvPTexp and pvPSPTexp. Nucleotide sequences of the 5′- and 3′-flanking regions, signal peptide, and terminator sequences of the phaseolin gene are shown; most of the coding sequence has been deleted to conserve space in the figure. The location and sequences of the six oligonucleotide primers used in CPCR-engineering both expression cassettes pvPTexp and pvPSPTexp are shown along with the restriction enzyme site sequences (internal *Nco* I or flanking *Hin*d III sites) synthesized within the CPCR oligonucleotide primers. Dashed lined regions of the oligonucleotide primers indicate shared identity with the coding or non-coding DNA strands. The oligonucleotide primers JLS-94, -133, and -96 were used to amplify the non-coding DNA strand (the DNA strand not shown). CCAAT, TATAAA, and polyadenylation signal (AATAAA) sequences are indicated by dashed lines above each sequence element. The phaseolin mRNA cap site is indicated by a # sign, located at position 425.

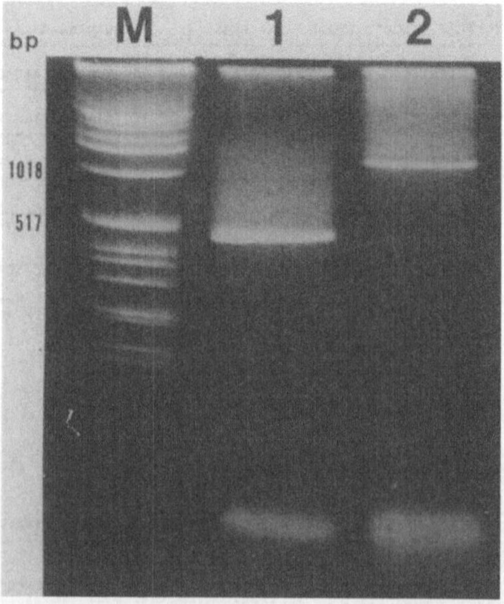

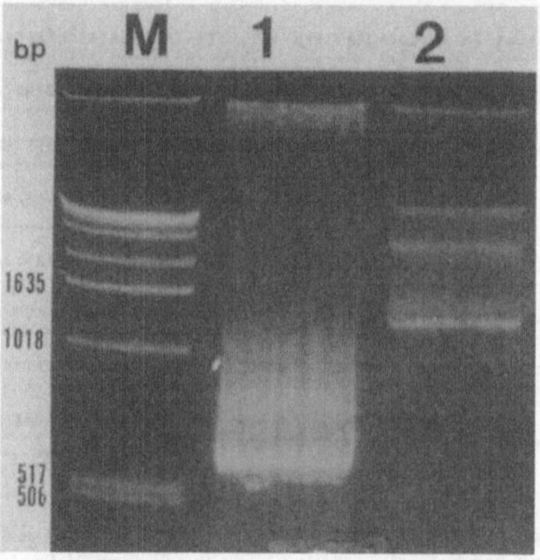

Fig. 6. CPCR amplification of promoter and terminator fragments for the phaseolin expression insert pvPTexp and pvPSPTexp. Frame A (top) shows CPCR amolification results for the 500 bp phaseolin promoter and 5′-untranslated fragment (lane 1) and 1170 bp phaseolin 3′-untranslated, polyadenylation signal, and 3′-flanking DNA regions (lane 2). Both CPCR fragments were digested with *Nco* I prior to loading onto the gel. The lane labeled M contains the 1 kb DNA size standard. The sequence and position of the oligonucleotide primers used for CPCR amplification are shown in Fig. 5. Frame B (bottom) shows CPCR amplification results for the 560 bp phaseolin 5′-flanking DNA region, 5′-untranslated, and signal peptide regions (lane 1) and the 1200 bp phaseolin termination, 3′-untranslated, polyadenylation signals, and 3′-flanking DNA region (lane 2). Both CPCR-amplified DNA fragments were digested with *Nco* I prior to loading onto the gel.

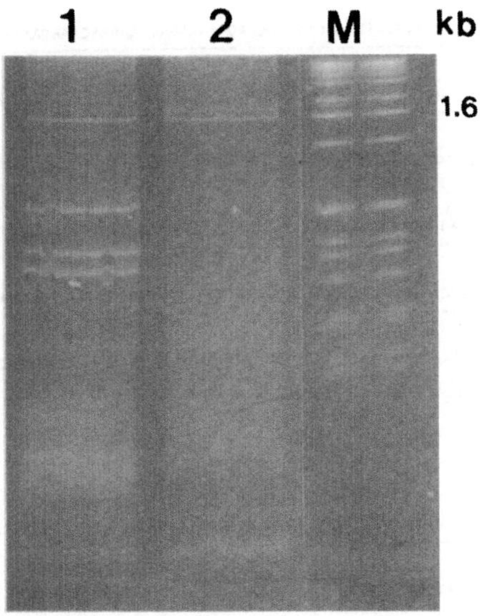

Fig. 7. The use of CPCR amplification to increase the yield of the ligation product pvPSPTexp. The phaseolin expression cassette pvPSPTexp fragment was assembled by ligating the *Nco* I-digested fragments shown in Fig. 6B and a small amount of the 1760 bp cassette fragment was obtained. One-half of the ligation product was purified by electrophoresis through a 7% acrylamide/15% glycerol gel (see Section 1, step 13). The other half of the ligation product was purified by phenol chloroform isoamyl alcohol extraction and EtOH-precipitated. Both fractions were subjected to a second CPCR amplification using oligonucleotide primers JLS-95 and -96. Both products were partially purified and then digested with *Hind* III followed by subsequent electrophoresis through a 7% acrylamide/15% glycerol gel. The lane labeled M contains the 1 kb DNA size standard. Lane 1 shows the amplification of the partially purified ligation fragment and lane 2 shows the amplification of the gel purified ligation fragment. In each lane, the band located above the 1.6 kb DNA size standard band in lane M corresponds to the expected 1790 bp fragment. These bands were removed and purified as described in Section 1, step 15. This DNA fragment was ligated directly into *Hind* III-digested and phosphatase treated pUC18 (see Section 1, step 17).

```
      HindIII
1     AAGCTTGGCACTCTGTGGTCTTTTGGTTCATGCATGGGTCTTGCGCAAGAAAAAGACAAAGAACAAAGAAAAAAGACAAAACAGAGAGACAAAACGCAAT
101   CACACAACCAACTCAAATTAGTCACTGGCTGATCAAGATCGCCGCGTCCATGTATGTCTAAATGCCATGCAAAGCAACACGTGCTTAACATGCACTTTAA
201   ATGGCTCACCCATCTCAACCCACACACAAACACATTGCCTTTTTCTTCATCATCACCACAACCACCTGTATATATTCATTCTCTTCCGCCACCTCAATTT
301   CTTCACTTCAACACACGTCAACCTGCATATGCGTGTCATCCCATGCCCAAATCTCCATGCATGTTCCAACCACCTTCTCTCTTATATAATACCTATAAAT
401   ACCTCTAATATCACTCACTTCTTTCATCATCCATCCATCCAGAGTACTACTACTCTACTACTATAATACCCCAACCCAACTCATATTCAATACTACTCTA
                                                                                                 NcoI
501   CTATGATGAGAGCAAGGGTTCCACTCCTGTTGCTGGGAATTCTTTTCCTGGCATCACTTTCTGCCTCATTTGCCACTTCACTCCGGCCATGGGCATTTGT
        MetMetArgAlaArgValProLeuLeuLeuLeuGlyIleLeuPheLeuAlaSerLeuSerAlaSerPheAlaThrSerLeuArgProTrpAlaPheVa
601   GTACTGAATAAGTATGAACTAAAATGCATGTAGGTGTAAGAGCTCATGGAGAGCATGGAATATTGTATCCGACCATGTAACAGTATAATAACTGAGCTCC
      lTyrEnd
701   ATCTCACTTCTTCTATGAATAAACAAAGGATGTTATGATATATTAACACTCTATCTATGCACCTTATTGTTCTATGATAAATTTCCTCTTATTATTATAA
801   ATCATCTGAATCGTGACGGCTTATGGAATGCTTCAAATAGTACAAAAACAAATGTGTACTATAAGACTTTCTAAACAATTCTAACTTTAGCATTGTGAAC
901   GAGACATAAGTGTTAAGAAGACATAACAATTATAATGGAAGAAGTTTGTCTCCATTTATATATTATATATTACCCACTTATGTATTATATTAGGATGTTA
1001  AGGAGACATAACAATTATAAAGAGAGAAGTTTGTATCCATTTATATATTATATACTACCCATTTATATATTATACTTATCCACTTATTTAATGTCTTTAT
1101  AAGGTTTGATCCATGATATTTCTAATATTTTAGTTGATATGTATATGAAAGGGTACTATTTGAACTCTCTTACTCTGTATAAAGGTTGGATCATCCTTAA
1201  AGTGGGTCTATTTAATTTTATTGCTTCTTACAGATAAAAAAAAAATTATGAGTTGGTTTGATAAAATATTGAAGGATTTAAAATAATAATAAATAATAAA
1301  TAACATATAATATATGTATATAAATTTATTATAATATAACATTTATCTATAAAAAAGTAAATATTGTCATAAATCTATACAATCGTTTAGCCTTGCTGGA
1401  CGACTCTCAATTATTTAAACGAGAGTAAACATATTTGACTTTTTGGTTATTTAACAAATTATTATTTAACACTATATGAAATTTTTTTTTTTTATCGGCA
1501  AGGAAATAAAATTAAATTAGGAGGGACAATGGTGTGTCCCAATCCTTATACAACCAACTTCCACAGGAAGGTCAGGTCGGGGACAACAAAAAAACAGGCA
1601  AGGGAAATTTTTTAATTTGGGTTGTCTTGTTTGCTGCATAATTTATGCAGTAAAACACTACACATAACCCTTTTAGCAGTAGAGCAATGGTTGACCGTGT
                                                                                    HindIII
1701  GCTTAGCTTCTTTTATTTTATTTTTTTATCAGCAAAGAATAAATAAAATAAAATGAGACACTTCAGGGATGTTTCAACCAAGCTT   1786
```

Fig. 8. Nucleotide sequence of the insert cloned into the phaseolin expression vector pUC18pvPSPTexpD22. The nucleotide sequence was obtained using double-stranded DNA and the dideoxy enzymatic sequencing method [48]. The 1 m DNA sequencing gel system described by Slightom *et al.* [52] is capable of obtaining nucleotide sequence readings in the range of 800 bp from sequencing primers. The phaseolin promoter and signal peptide coding sequences were linked by a *Nco* I site to the phaseolin terminator, 3′-untranslated, polyadenylation signal, and 3′-flanking DNA region. The flanking CPCR-added *Hin*d III sites flanking pvPSPTexp are also indicated (the integrity of these sites was confirmed by sequencing from oligonucleotide primers made to sequences flanking the pUC18 polylinker. The phaseolin gene regulatory elements, CCAAT, TATAAA, and AATAAA, are indicated with dashed lines above their sequences. The site of translation initiation is denoted by an #. The sequences of 22 to 25 amino acids phaseolin signal peptide [51] and the four amino acids adjacent to the translation termination codon are shown.

Section 4. CPCR engineering of a seed-specific promoter cassette directly from genomic DNA

The purpose of this section is to demonstrate further the power of the CPCR-engineering technique by using it to engineer the promoter portion of a seed-specific expression cassette using total genomic DNA as the source of the target sequences. For this example we CPCR-engineered the promoter region of the seed storage protein gene that encodes the α' subunit of β-conglycinin by using total *Glycine max* (soybean) genomic DNA. The requirements for this experiment are simple: the sequence of the selected promoter region and a small amount of soybean DNA must be available. The sequence of the gene encoding the α' subunit of β-conglycinin was described by Doyle *et al.* [15]. This description also includes about 800 bp of its 5′-flanking DNA sequence, which contains the DNA elements responsible for tissue specificity and developmental regulation [11]. Chen *et al.* [11] showed that most of these important sequences elements are located within about 300 bp 5′ of the transcriptional start site. To ensure inclusion of these sequences, the oligonucleotide primers were designed to amplify a region of 620 bp, which includes 550 bp of 5′-flanking DNA (regulatory DNA), the 57 bp 5′-untranslated region, and places a *Nco* I site at the translation initiation codon. This 620 bp CPCR-engineered promoter fragment of the α' subunit of β-conglycinin gene is referred to as 'gmP'. It is not necessary to CPCR-engineer the 3′-regulatory region of this soybean gene because the analogous region of the phaseolin gene (Section 3, step 3, fragment b) can be used to complete the expression cassette. This complete expression cassette fragment is referred to as gmPpvTexp, and it can be used to express beneficial gene-coding regions that contain their own signal peptides and termination codons similar to the phaseolin expression vector pUC18pvPTexp (see Section 3).

4.1. CPCR engineering the α' subunit of β-conglycinin gene promoter

Steps in the procedure

1. From the nucleotide sequence of the target gene, design oligonucleotide primers that will amplify the promoter region. In many cases the location of important DNA elements that regulate expression will not be known; in these cases the inclusion of 1 kb (if the sequence is available) of the 5′-flanking region of a gene should be sufficient, as these sequence elements are generally located within 500 bp 5′ of the transcription initiation site [7]. For this example the location of these important sequences is

known; Chen *et al.* [11] suggests that these regulatory sequences are located within 300 bp 5′ of the transcription initiation site (see Fig. 12). The following oligonucleotide primers were synthesized:

– JLS-97 (5′-TACATGAAGCTTGATCCATGCCTTCATTTGCCG-3′); this primer adds a *Hind* III site 550 bp 5′ of the transcription initiation site (Fig. 9);

– JLS-98 (5′-GTCAAGCCATGGTTAGTATATCTTAAATTCTTTAATAC-3′); this primer adds a *Nco* I site at the translation initiation site (Fig. 9); this CPCR-engineering step also removes the potential redundant translation initiation sites found in this gene [15] and replaces these sequences with one translation initiation site that conforms to the plant consensus [26, 27].

2. Obtain total genomic soybean DNA using any procedure that yields reasonably pure DNA [34, 44, 45, and Chee *et al.*, Chapter C3] and determine its concentration. Also determine the concentration of the oligonucleotide primers (Section 1, step 4). Use about 1 μg of total soybean DNA and 600 ng of each oligonucleotide primer. For this example, the soybean DNA and the two oligonucleotide primers had a concentration of 500 μg/ml and 200 μg/ml, respectively. The CPCR components should be added in the following order:

Component	*Order of addition*	*Volume* (μl)
Double-distilled sterile H_2O	1	65.5
PCR buffer (10×) (Perkin-Elmer-Cetus)	2	10.0
dNTP mix, 1.25 mM (each dNTP) (Perkin-Elmer-Cetus)	3	16.0
Primer JLS-97 (200 μg/ml)	4	3.0
Primer JLS-98 (200 μg/ml)	5	3.0
Genomic soybean DNA (500 μg/ml)	6	2.0
Taq polymerase (2.5 units/μl) (Perkin-Elmer-Cetus)	7	0.5
	Total	100.0

3. Mix gently, then briefly centrifuge. Overlay the sample with 100 μl of mineral oil to prevent evaporation.
4. CPCR cycle temperatures were controlled by an automated temperature cycler from Ericomp, Inc. The following temperature cycles were used:

Duplex denaturation	94 °C	2 min
Annealing	55 °C	2 min
DNA synthesis	72 °C	3 min

5. The automated cycler repeated this temperature cycles thirty times and on the final cycle the duplex denaturation step was omitted and the DNA synthesis step was extended to 10 min to aid the complete synthesis of the accumulated DNAs.
6. The CPCR-amplified gmP product was analyzed as described in Section 1 (step 10) and this result is shown in Fig. 10. A sufficient amount of the 620 bp gmP product was available and it was prepared for ligation to the phaseolin terminator fragment following the procedures described in Section 1 (steps 11 to 13). The gmP and pvT ligation was done as described in Section 1 (step 14), and ligation product was purified (the expected 1790 bp gmPpvT fragment) as described in Section 1 (step 15). This gmPpvT fragment was digested with *Hind* III before gel purifying. The ligated gmPpvT fragment was cloned into the *Hind* III-digested pUC18 vector prepared in Section 1 (step 17), this ligation was done as described in Section 1 (step 18).
7. About 3 μl of this ligation product was used to transform 100 μl of competent *E. coli* cells, and the transformants were selected by plating on AB agar plates containing 100 μg/ml ampicillin. A total of 30 colonies were obtained.
8. PCR amplification was used to screen these clones. Crude plasmid DNAs were prepared as described by Morelle [32], and this DNA was used as template for PCR amplifications. The oligonucleotide primers (JLS-97 and -98) were used to amplify the gmP fragment and oligonucleotide primers (JLS-97 and -96) were used to amplify the complete insert gmPpvT. The results of this PCR analysis are shown in Fig. 11.

Notes

8. The two positive clones identified by PCR analysis (Fig. 11) were subjected to restriction enzyme site mapping and found to contain the expected 1790 bp *Hind* III fragment, and a *Hind* III and *Nco* I double digest yielded the expected 620 bp gmP and 1170 bp pvT fragments. The inserts from clones pUC18gmPpvTexp102 and 104 were completely sequenced to determine if they contained any ligation and/or PCR-generated errors. The nucleotide sequence of the two CPCR-engineered gmP promoter regions are shown in Fig. 12. These sequences show that the promoter region of the soybean seed storage protein gene α' subunit of β-conglycinin can be obtained directly from total soybean genomic DNA. The sequences presented in Fig. 12 do not completely match each other nor the sequence reported by Doyle *et al.* [15]. The possibility exists that these differences

are PCR-generated errors because these clones were obtained from total genomic DNA. The initial amount of template sequences is considerably less than that used in the experiments described in Sections 1 to 3. We can not presently determine if any of these nucleotide differences are truly the result of PCR-generated errors because the α' subunit of β-conglycinin storage protein is encoded in a multigene family [15]. Thus, the amplified gmP sequences presented in Fig. 12 could have been derived from another member of this gene family, whose sequence differs slightly from that published by Doyle *et al.* [15]. The degree of variation found among these CPCR promoters is only about 1%, which is similar to the level expected for members of the same gene family (J.L.Slightom and R.F.Drong, unpublished data). In addition, different sources of soybean genomic DNAs were used; Doyle *et al.* [15] used DNA isolated from the soybean cultivar 'Wayne' and we isolated DNA from the 'Asgrow' cultivar A3733.

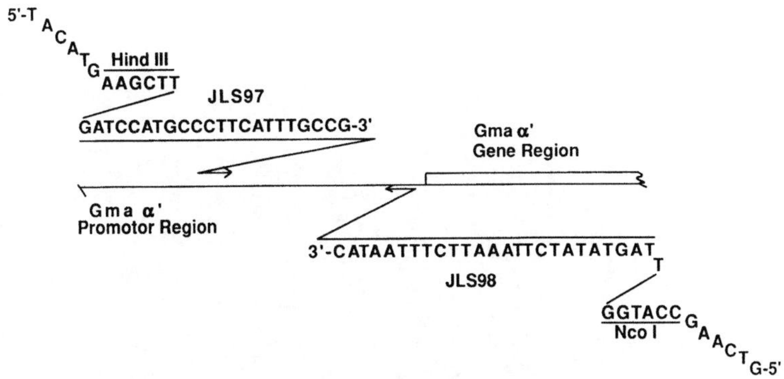

Fig. 9. Diagram of the CPCR strategy for engineering the promoter of the soybean gene encoding the α' subunit of β-conglycinin of *Glycine max.* A schematic diagram of the promotor and partial gene region of the α' subunit of β-conglycinin shows the location, direction, and sequence of the two oligonucleotide primers used to CPCR-engineer the expression cassette fragment gmP. The oligonucleotide primer JLS-97 contains the restriction enzyme recognition site for *Hind* III, while JLS-98 contains the recognition site for *Nco* I. Both oligonucleotide primers contain 6 additional 5′ nucleotides used to aid restriction enzyme digestion.

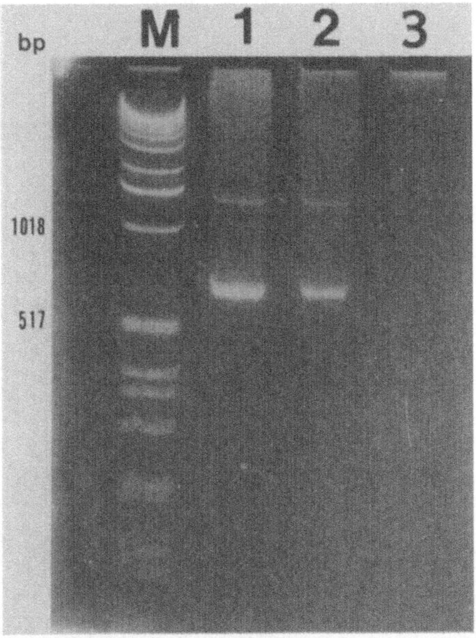

Fig. 10. CPCR amplification of the promoter region gmP using total soybean DNA as the source of target gene sequences. Total genomic DNA from two cultivars of soybean, Asgrow A3733 and Wayne, was isolated [34, 44, 45] and subjected to CPCR-amplification with oligonucleotide primers JLS-97 and -98 (see Section 4). The lane labeled M contains the 1 kb DNA size standard. Lane 1 shows the presence of a 600 bp CPCR amplification DNA fragment (*Nco* I digested) obtained using total soybean DNA isolated from Wayne and lane 2 shows that similar results using total DNA isolated from the Asgrow soybean cultivar, A3733. Lane 3 contains negative control results where oligonucleotide primers JLS-97 and -98 were used for CPCR amplification of maize genomic DNA; thus no visible DNA band was obtained.

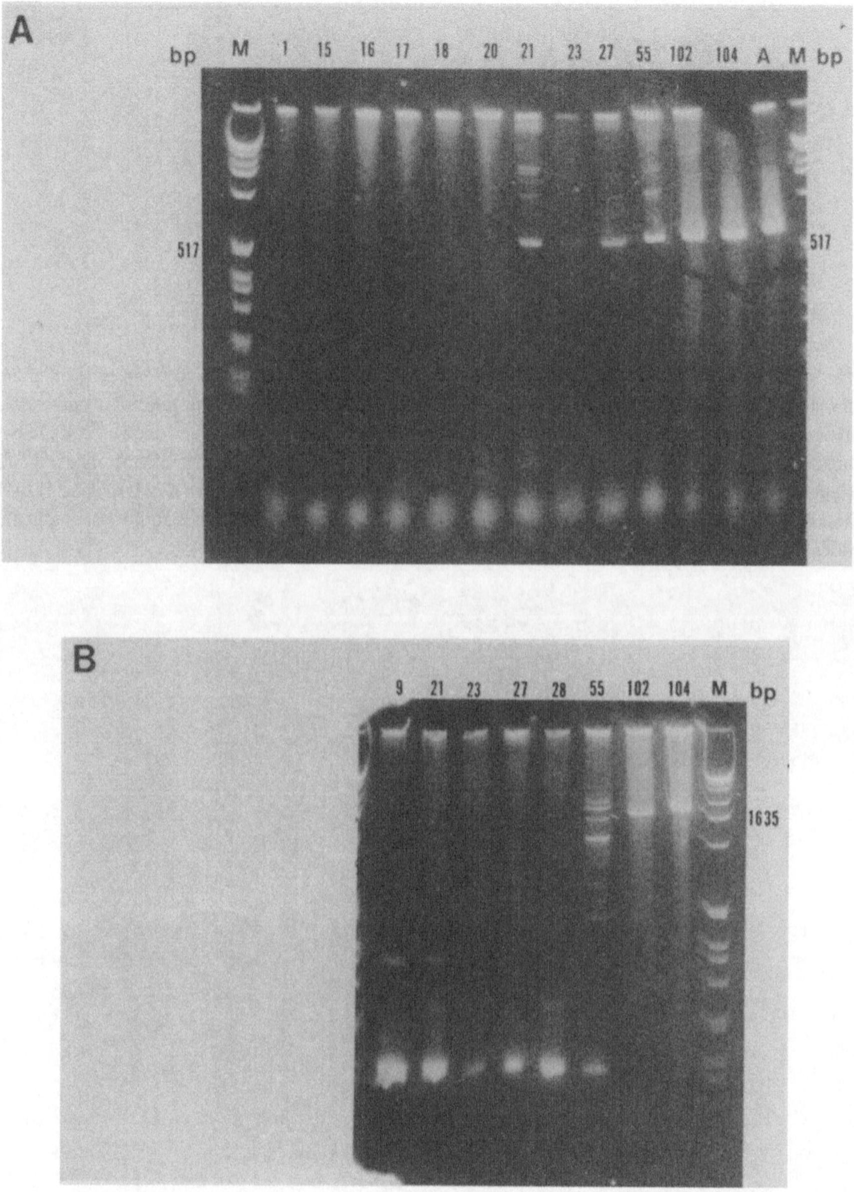

Fig. 11. PCR screening of putative clones that may contain the correct gmPpvTexp insert. Crude plasmid DNAs were isolated from *E. coli* clones using the procedure described by Morelle (1989), and they were screened by PCR amplification by adding either oligonucleotide primers JLS-97 and -98 (which should yield the 620 bp promoter fragment) or oligonucleotide primers JLS-97 and -96 (which should yield the 1790 bp that represents the complete cloned DNA fragment). The lanes are labeled according to the clone sample number. Frame A: PCR analysis using oligonucleotide primers JLS-97 and -98 on cloned DNAs whose numbers are indicated at the tops of each lane. This analysis shows that clones 21, 27, 55, 102 and 104 contain the expected 620 bp gmP fragment; these clones

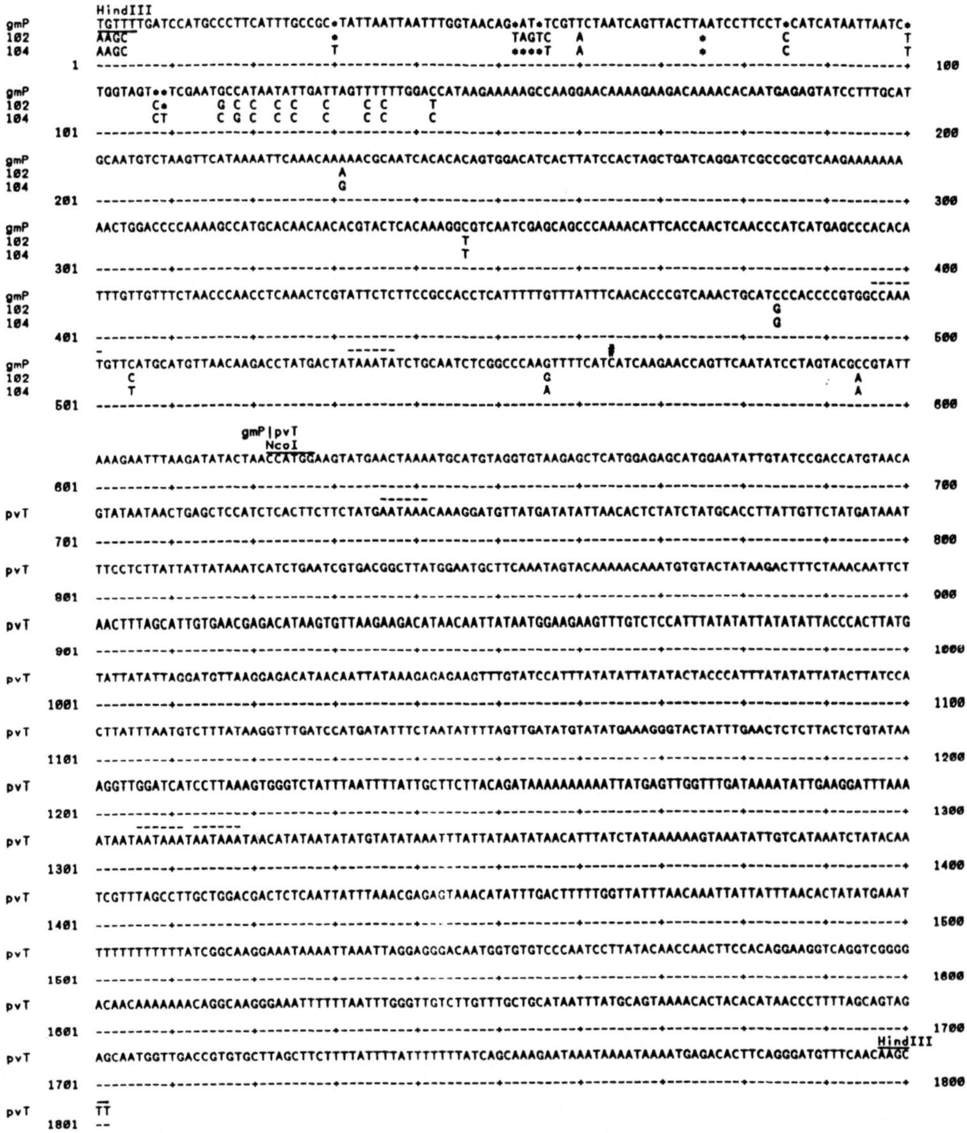

Fig. 12. Nucleotide sequence results for the CPCR-engineered soybean α′ subunit of β-conglycinin gene promoter fused to the phaseolin terminator, gmPpvT. The nucleotide sequence was obtained using the same procedure as described in the legend to Fig. 8. The nucleotide sequence of the cloned

---

were investigated further. The lane labeled A contains the gmP fragment shown in Fig. 10, which served as a positive control size marker. Frame B: PCR analysis using oligonucleotide primers JLS-97 and -96 on the clones that were positive in frame A. The expected 1790 bp fragment was found in only clone numbers 102 and 104; both of these clones were investigated further by obtaining their nucleotide sequence (see Fig. 12). In both frames, the lanes labeled M contain the 1 kb DNA size standard.

gene reported by Doyle *et al.* [15] is shown on the top line, while only variant nucleotide sequence positions are shown for the gmP sequences from clones 102 and 104. Gaps, used to maximize nucleotide sequence identities among these sequences, are designated by asterisks. The location of the junction between the gmP promoter and the phaseolin fragment pvT is indicated above the CPCR-added *Nco* I site. Flanking CPCR-added *Hin*d III sites of gmPpvTexp are indicated. Nucleotide sequence elements, CCAAT, TATAAA, and AATAAA, are denoted by a dashed line above their sequences. The transcription initiation site is denoted by the # sign.

Section 5. The use of CPCR to purify and modify synthesized oligonucleotides

The synthesis of complete genes has become possible with the advent of efficient methods for the chemical synthesis of oligodeoxynucleotides [5, 49]. The construction of a complete gene usually involves the synthesis of many oligonucleotides ranging in size from 30 to 80 nucleotides with overlaps of about 6 bp at each junction. These oligonucleotides are annealed, repaired using *E. coli* polymerase (Klenow fragment), and ligated using T4 DNA ligase. The resulting double-stranded fragment can then be cloned [30]. This technique has been widely used for the synthesis of a number of genes, some of them quite large (such as human t-PA [19,42]).

Nucleotide syntheses in the range of 80 nucleotides are common and the percent of full-length oligonucleotides is generally in the range of 30 to 40%. However, if the synthesis of oligonucleotides is extended in the range of 130 bp the yield of full-length oligonucleotides is greatly reduced, ranging between 1 to 10% depending on reaction efficiencies. The major problem with the synthesis of such a long oligonucleotide is that the small percentage of full-length oligonucleotide is heavily contaminated with reaction by-products and pre-terminated oligonucleotides making purification by standard methods (acrylamide gel, HPLC, OPC cartridge) difficult or impossible. In this section we demonstrate how CPCR engineering can be used to aid in the isolation and modification of a 144 nucleotide oligonucleotide. This oligonucleotide was designed to encode a major portion of the Brazil nut gene, a 2S albumin seed protein gene of the Brazil nut tree (*Bertholletia excelsa*).

Our interest in synthesizing this particular 144 nucleotide oligonucleotide is in its expression (using the expression cassette designed in Section 3, pUC18pvPSPTexp) in the seeds of legumes such as soybean (*Glycine max*), pea (*Pisum sativum*), and common bean (*Phaseolus vulgaris*). Such expression would have the potential of increasing the sulfur-containing amino acid content present in these seed proteins. A long anticipated goal of plant molecular biology research has been the use of genetic engineering techniques to correct the sulfur amino acid deficiencies in the seeds of these plant species.

The 2S albumin seed storage protein gene of the Brazil nut tree has been cloned and sequenced by Altenbach *et al.* [2], and this gene encodes a polypeptide that undergoes a series of proteolytic cleavages that results in a final 12 kDa protein composed of a 3 kDa and 9 kDa peptide held together by disulfide bonds. The Brazil nut gene has been expressed in tobacco after fusion to the phaseolin seed storage protein gene promoter [3]. In some transgenic tobacco seeds as much as 8% of the total seed protein is the 2S Brazil nut protein, which results in a 30% increase in the sulfur content of these seeds. These results are extremely encouraging as they show that the Brazil nut protein can be expressed and stably accumulated within the seeds of a foreign plant species.

Synthesis of the complete Brazil nut gene coding region would require a total of 504 bp. However, complete synthesis of the gene is not necessary because much of the peptide is degraded and thus does not accumulate in the seed [2]. In addition, the expression cassette pUC18pvPSPTexp contains the necessary transcription and translation regulatory elements (signal peptide, translation initiation and termination codons, and polyadenylation signals). Thus, only the gene region that encodes the sulfur-containing amino acid residues needs to be synthesized. This region of sulfur-containing amino acid residues is contained within the 9 kDa peptide where out of 77 amino acids, 14 residues are methionine (18%) and six residues are cysteine (8%). This region can be shortened even more because many of these sulfur-containing amino acids are clustered, one such cluster contains five Met residues in a stretch of six residues. We describe here the initial step in the synthesis of the 9 kDa subunit, the synthesis of a stretch of 144 nucleotides (48 amino acids) which encodes a polypeptide with a sulfur amino acid content of 35%. This 144 nucleotide region was synthesized as a single oligonucleotide (Fig. 13) on an Applied Biosystems DNA synthesizer (model 380B). This 144 nucleotide oligonucleotide is referred to as BZ-1.

4.1. CPCR amplification to purify and engineer BZ-1 oligonucleotide

Steps in the procedure

1. Synthesis of a long oligonucleotide should be monitored by collecting the trityl byproduct and reading its OD_{447} [49]. The final trityl reading for the synthesis of BZ-1 indicated that about 9% of the oligonucleotides were full-length. After deprotection (60 °C in NH_4OH), the NH_4OH was removed by vacuum in a Savant Speedvac, and the dry sample was resuspended in 100 μl of TEN buffer. The nucleic acid concentration was determined (OD_{260} = 500); however, most of this absorption represents pre-terminated oligonucleotides. Assuming 20 μg per OD unit the concentration of the sample is 10 mg/ml. For convenience, part of this sample was diluted 100-fold to a final concentration of 100 μg/ml.
2. The trityl readings indicate that the full-length BZ-1 oligonucleotide is heavily contaminated. Its purification is a difficult problem; the OPC car-

tridges described in Section 1 are not effective for the purification of oligonucleotides longer than about 70 bp. We attempted to purify the oligonucleotide by electrophoresis through a 7% acrylamide-7 M urea gel. Figure 14 shows the results of this purification attempt in which 1 to 8 μg of BZ-1 was loaded. If any full-length BZ-1 oligonucleotide is present it is so heavily contaminated that it is difficult to determine its location.

3. To determine if any full-length BZ-1 oligonucleotide was present we used CPCR amplification using two oligonucleotide primers, JLS-136 and -137 (Fig. 13), which share identity with the 5′ and 3′ ends of BZ-1, respectively. These oligonucleotide primers also add flanking *Nco* I sites for later cloning. CPCR amplification of the BZ-1 oligonucleotide with oligonucleotide primers JLS-136 and -137 would provide the complementary DNA strand and add 24 bp to the amplified DNA product (12 bp encoding the two *Nco* I sites and the extra 12 bp added to aid enzyme digestion), thus the expected CPCR DNA fragment should be about 170 bp in length.
4. Because the concentration of the full-length BZ-1 oligonucleotide is unknown and the effect of the contaminants in the mixture on the CPCR reaction is also unknown, a series CPCR amplifications were used. These reactions contained between 70 ng to 1.2 μg of the crude BZ-1 mixture. These CPCR samples were set-up as described in Sections 1 (steps 7 to 9) and the CPCR results were analyzed as described in Section 1 (step 10). Interestingly, only the sample that contained the least amount of BZ-1 template (70 ng) showed the presence of the expected 170 bp DNA band.
5. Additional CPCR samples containing 35 or 70 ng of the crude BZ-1 oligonucleotide subjected to electrophoresis to obtain more of the CPCR-engineered BZ-1 fragment. As shown in Fig. 15, all of these CPCR samples yielded the expected size BZ-1 fragment, indicating that the optimum amount of the crude BZ-1 sample to be used was in the range of 70 ng.
6. The 170 bp BZ-1 fragment was removed from the acrylamide gel as described in Section 1 (steps 12 and 13), and to increase the amount of BZ-1 fragment a fraction of this purified sample was subjected to a second round of CPCR amplification (see Section 1, step 16). After determining if the second CPCR amplification product yielded a sufficient amount of the fragment (Section 1, step 10) the *Taq* polymerase was removed and the DNA was digested with *Nco* I (10 units/μg of *Nco* I and incubated at 37 °C for 2 h), see Section 1 (step 11). The restriction enzyme was removed by extraction with the phenol : chloroform : isoamyl alcohol mixture followed by purification using a 7% acrylamide/15% glycerol gel (Section 1, steps

12 and 13). The ethidium bromide staining pattern of this gel is shown in Fig. 16.

7. We cloned the purified *Nco*I-cut BZ-1 fragments into a *Nco*I site of pUC18cpexp, which is described in Section 1 (it was available at this time; see Section 2). The ligation reaction was set-up as described in Section 1 (step 18) and about 3 μl of this ligation product was used to transform 100 μl of competent *E. coli* cells. This transformation yielded about 500 colonies, of which 48 were selected for restriction enzyme site mapping. The results of these digestion for clones number 25 to 48 are shown in Fig. 17, which indicates that the full-length BZ-1 fragment can be recovered at high frequency.
8. Nucleotide sequence analysis of the BZ-1 insert in clones pUC18cpBZ-1-25, -26, -28, -29, and -38 were determined. All of the BZ-1 inserts contained the expected sequence; no DNA synthesis, PCR, and/or ligation errors were found and only clone 38 was abnormal as it contained a second (truncated) BZ-1 insert (see Fig. 12).
9. The BZ-1 insert from clone pUC18cpBZ-1-25 was subsequently transferred into the seed specific expression vector pUC18pvPSPTexp. The sequence of the clone referred to as pUC18pvPSPTBZ-1 is shown in Fig. 18.

Notes

1. If this reading drops off sharply, the synthesis cycle may have been interrupted and should be re-initiated.
4. Obtaining the expected 170 bp CPCR fragment was encouraging because it indicated that the crude BZ-1 sample does contain some full-length or nearly full-length BZ-1 molecules (nearly full-length BZ-1 oligonucleotides could be amplified if they overlap with the oligonucleotide primers JLS-136 and -137; see Fig. 13). However, this result also suggests that contaminants in this crude sample can inhibit some aspect of PCR, possibly by interfering with template annealing or the functioning of *Taq* polymerase.
6. The presence of two distinct DNA bands suggests that about half of the BZ-1 fragments was cut (represented in the lower 150 bp band), while the other half remained uncut (represented in the upper 170 bp band). The cut and uncut BZ-1 fragments were not well separated (Fig. 16) and it was difficult to purify the cut fragments from the gel matrix without contaminating it with some uncut BZ-1 fragments. This was not a serious problem as the uncut BZ-1 fragments should not interfere with the ligation step because the primers used do not contain 5′ acceptor phosphate groups. They thus cannot be ligated to themselves or into the phosphatase treated cloning vector.
7. Fig. 17 also shows that the BZ-1 DNA fragment isolation from the gel shown in Fig. 16 also contains a large number of abnormal BZ-1 fragments. Several clones show the presence of more than one BZ-1 (or derivative) fragments.
8. The example described here extends the usefulness of CPCR amplification by using it to purify, complement, and modify synthesized DNAs. We have demonstrated that, at the

present reaction efficiency of the Applied Biosystems DNA synthesizer (98.5 to 99.5% per cycle), we can use CPCR amplification to purify a 144 nucleotide oligonucleotide even though its concentration in the final reaction mixture is less than 10% (as Fig. 14 shows its concentration appears to be much less). If oligonucleotides in the range of 150 bp can be routinely synthesized (with purifications, enrichments, and modifications being provided by the CPCR amplification technique), the synthesis of complete genes could be accomplished with far fewer oligonucleotides and ligation steps. In addition, as the efficiency of the oligonucleotide synthesis reactions improves (99.9% and better) the combined use of CPCR should allow one step synthesis of oligonucleotides to be extended to the 300 to 500 nucleotide range. The ability to synthesize and use oligonucleotides of such lengths will be a great benefit to basic molecular biology research and to applied research areas that require genetic engineering techniques.

Acknowledgments

We wish to thank Nicole Hatzenbuhler and Mary Shea for the synthesis of oligomer primers used for CPCR amplification and nucleotide sequencing.

```
                    NcoI
                  ------
JLS-136  AATTATCCATGG -----------------------
                      ATGGAGCCGCACATGAGCGAGTGCTGCGAGCAGCTGGAGGGGATGGACGAGAGCTGCAGA
                    1 ---------+---------+---------+---------+---------+---------+  60
                      MetGluProHisMetSerGluCysCysGluGlnLeuGluGlyMetAspGluSerCysArg

                      TGCGAAGGCTTAAGGATGATGATGATGATGAGGATGCAACAGGAGGAGATGCAACCCCGA
                   61 ---------+---------+---------+---------+---------+---------+  120
                      CysGluGlyLeuArgMetMetMetMetMetArgMetGlnGlnGluGluMetGlnProArg

                      GGGGAGCAGATGCGAAGGATGATG
                  121 ---------+---------+---- 144
                         CTCGTCTACGCTTCCTACTACGGTACCTCTTAA    JLS-137
                      GlyGluGlnMetArgArgMetMet
                                               ------
                                                NcoI
```

Fig. 13. Nucleotide sequence of the region of the Brazil nut gene 2S albumin seed storage protein gene that encodes a large number of sulfur-containing amino acids. The sequence is from Altenbach *et al.* [2]. The sequence shown here represents 144 bp of the 9 kDa subunit of this gene (note the sequence shown here contains one additional Met residue not found in the native Brazil nut gene). The oligonucleotide primers, JLS-136 and -137, were used for CPCR amplification of the 144 nucleotide BZ-1 oligonucleotide and to add the flanking *Nco* I restriction enzyme recognition sites.

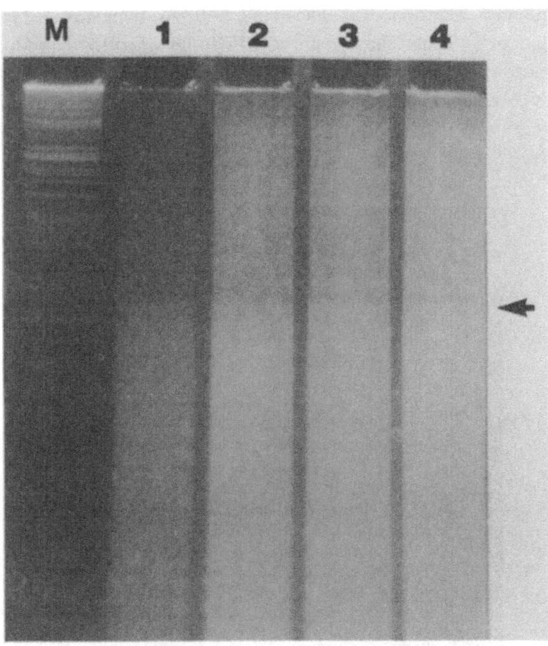

Fig. 14. Analysis of the synthesized 144 nucleotide oligomer designated BZ-1. Various amounts of this oligonucleotide were added to DNA loading buffer (80% formamide, 10 mM NaOH, 1.0 mM EDTA, 0.1% xylene cyanol), heated to 90 °C for 3 min, and loaded onto a 7% acrylamide 7 M urea sequencing gel. The following amounts of the BZ-1 oligonucleotide were added; lane 1, 1 μg; lane 2, 2 μg; lane 3, 4 μg; lane 4, 8 μg. A DNA size standard, the 1 kb DNA size standard, is in the lane labeled M; however, because the gel is a denaturing gel, it does not indicate corrected DNA sizes. The arrow indicates the possible position of the 144 base oligonucleotide BZ-1.

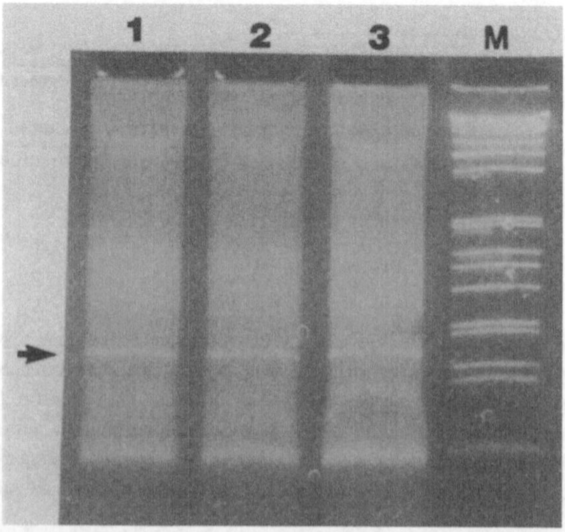

Fig. 15. CPCR amplification of BZ-1 and analysis of amplified products. Three separate CPCR samples were subjected to electrophoresis. The amount of BZ-1 template used is as follows: lane 1, 35 ng; lane 2, 35 ng; lane 3, 70 ng. The amount of CPCR-amplified BZ-1 product obtained for all three reactions appear to be about the same. The lane labeled M contains the 1 kb DNA size standard. The arrow points to the presumed CPCR-amplified BZ-1 fragment (double stranded) that is about 170 bp in size, due to the addition of extra nucleotides associated with oligonucleotides JLS-136 and -137 (see Section 5, step 3).

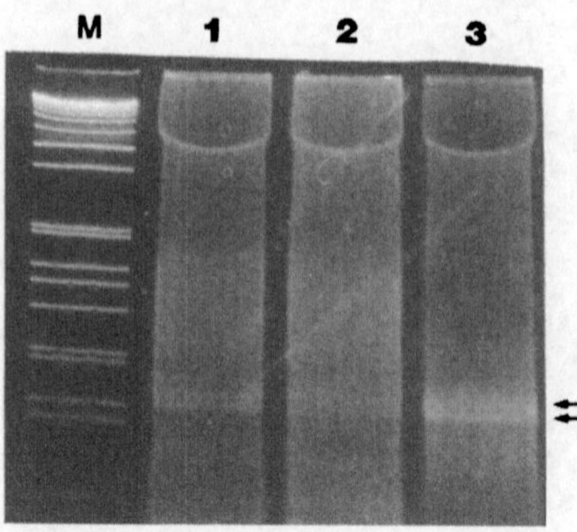

Fig. 16. Nco I digestion of BZ-1 fragments obtained after a second CPCR amplification. About 10% of the DNA fragment isolated from the respective lanes shown in Fig. 15 were subjected to a second CPCR amplification, followed by digestion with *Nco* I. Electrophoresis of these samples through a 7% acrylamide/15% glycerol gel shows two fragments (see arrow) The upper fragment corresponds in size to the uncut BZ-1 fragment (about 170 bp, see Fig. 15) and the lower fragment corresponds in size to the *Nco* I-cut BZ-1 fragment (about 150 bp). The lane labeled M contains the 1 kb DNA size standard.

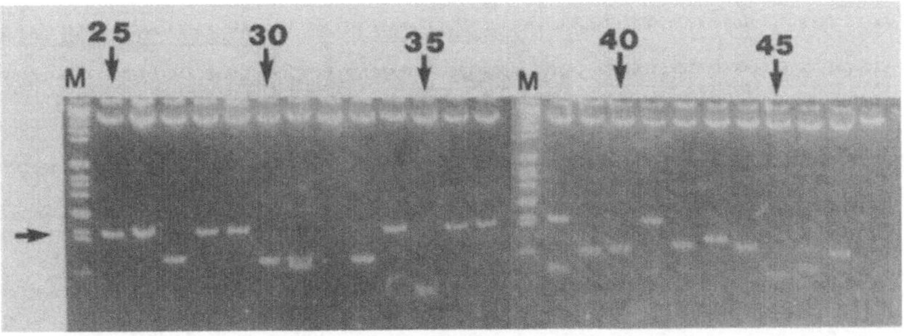

Fig. 17. Restriction enzyme site analysis of clones resulting for the ligation of the BZ-1 fragment and the expression vector pUC18cpexp. DNA was isolated from bacterial clones using the method described by Morelle [32], digested with *Nco* I, and subjected to electrophoresis through a 7% acrylamide/15% glycerol gel. The first 24 clones analyzed did not contain any inserts near the expected size and thus are not shown; however, a similar analysis of clones 25 to 48 showed several that contain inserts of the expected size, see clone numbers 25, 26, 28, 29, 34, 36, 37, 38, and 41. Clone 38 appears to contain two inserts. The lanes labeled M contain the 1 kb DNA size standard.

```
      HindIII
      ------
      AAGCTTGGCACTCTGTGGTCTTTTGGTTCATGCATGGGTCTTGCGCAAGAAAAAGACAAAGAACAAAGAAAAAAGACAAAACAGAGAGACAAAACGCAAT
1     ---------+---------+---------+---------+---------+---------+---------+---------+---------+---------+  100

      CACACAACCAACTCAAATTAGTCACTGGCTGATCAAGATCGCCGCGTCCATGTATGTCTAAATGCCATGCAAAGCAACACGTGCTTAACATGCACTTTAA
101   ---------+---------+---------+---------+---------+---------+---------+---------+---------+---------+  200

      ATGGCTCACCCATCTCAACCCACACACAAACACATTGCCTTTTTCTTCATCATCACCACAACCACCTGTATATATTCATTCTCTTCCGCCACCTCAATTT
201   ---------+---------+---------+---------+---------+---------+---------+---------+---------+---------+  300
                                                   -----                                            -----
      CTTCACTTCAACACACGTCAACCTGCATATGCGTGTCATCCCATGCCCAAATCTCCATGCATGTTCCAACCACCTTCTCTCTTATATAATACCTATAAAT
301   ---------+---------+---------+---------+---------+---------+---------+---------+---------+---------+  400
                                 #
      ACCTCTAATATCACTCACTTCTTTCATCATCCATCCATCCAGAGTACTACTACTCTACTACTATAATACCCCAACCCAACTCATATTCAATACTACTCTA
401   ---------+---------+---------+---------+---------+---------+---------+---------+---------+---------+  500

                                                                                               NcoI
                                                                                               ------
      CTATGATGAGAGCAAGGGTTCCACTCCTGTTGCTGGGAATTCTTTTCCTGGCATCACTTTCTGCCTCATTTGCCACTTCACTCCGGCCATGGATGGAGCC
501   ---------+---------+---------+---------+---------+---------+---------+---------+---------+---------+  600
        M  M  R  A  R  V  P  L  L  L  L  G  I  L  F  L  A  S  L  S  A  S  F  A  T  S  L  R  P  W  M  E  P
                                                                             ↑  ↑  ↑  ↑
                                                                      Signal Peptide Cleavage Sites
      GCACATGAGCGAGTGCTGCGAGCAGCTGGAGGGGATGGACGAGAGCTGCAGATGCGAAGGCTTAAGGATGATGATGATGATGAGGATGCAACAGGAGGAG
601   ---------+---------+---------+---------+---------+---------+---------+---------+---------+---------+  700
       H  M  S  E  C  C  E  Q  L  E  G  M  D  E  S  C  R  C  E  G  L  R  M  M  M  M  M  R  M  Q  Q  E  E
                                          NcoI
                                          ------
      ATGCAACCCCGAGGGGAGCAGATGCGAAGGATGATGCCATGGGCATTTGTGTACTGAATAAGTATGAACTAAAATGCATGTAGGTGTAAGAGCTCATGGA
701   ---------+---------+---------+---------+---------+---------+---------+---------+---------+---------+  800
      M  Q  P  R  G  E  Q  M  R  R  M  M  P  W  A  F  V  Y  *
                                                                                    Poly A
                                                                                    ------
      GAGCATGGAATATTGTATCCGACCATGTAACAGTATAATAACTGAGCTCCATCTCACTTCTTCTATGAATAAACAAAGGATGTTATGATATATTAACACT
801   ---------+---------+---------+---------+---------+---------+---------+---------+---------+---------+  900

      CTATCTATGCACCTTATTGTTCTATGATAAATTTCCTCTTATTATTATAAATCATCTGAATCGTGACGGCTTATGGAATGCTTCAAATAGTACAAAAACA
901   ---------+---------+---------+---------+---------+---------+---------+---------+---------+---------+  1000

      AATGTGTACTATAAGACTTTCTAAACAATTCTAACTTTAGCATTGTGAACGAGACATAAGTGTTAAGAAGACATAACAATTATAATGGAAGAAGTTTGTC
1001  ---------+---------+---------+---------+---------+---------+---------+---------+---------+---------+  1100

      TCCATTTATATATTATATATTACCCACTTATGTATTATATTAGGATGTTAAGGAGACATAACAATTATAAAGAGAGAAGTTTGTATCCATTTATATATTA
1101  ---------+---------+---------+---------+---------+---------+---------+---------+---------+---------+  1200

      TATACTACCCATTTATATATTATACTTATCCACTTATTTAATGTCTTTATAAGGTTTGATCCATGATATTTCTAATATTTTAGTTGATATGTATATGAAA
1201  ---------+---------+---------+---------+---------+---------+---------+---------+---------+---------+  1300

      GGGTACTATTTGAACTCTCTTACTCTGTATAAAGGTTGGATCATCCTTAAAGTGGGTCTATTTAATTTTATTGCTTCTTACAGATAAAAAAAAAATTATG
1301  ---------+---------+---------+---------+---------+---------+---------+---------+---------+---------+  1400

                                      Poly A type sequences
                                        ----------------
      AGTTGGTTTGATAAAATATTGAAGGATTTAAAATAATAATAAATAATAAATAACATATAATATATGTATATAAATTTATTATAATATAACATTTATCTAT
1401  ---------+---------+---------+---------+---------+---------+---------+---------+---------+---------+  1500

      AAAAAAGTAAATATTGTCATAAATCTATACAATCGTTTAGCCTTGCTGGACGACTCTCAATTATTTAAACGAGAGTAAACATATTTGACTTTTTGGTTAT
1501  ---------+---------+---------+---------+---------+---------+---------+---------+---------+---------+  1600

      TTAACAAATTATTATTTAACACTATATGAAATTTTTTTTTTTTTATCGGCAAGGAAATAAAATTAAATTAGGAGGGACAATGGTGTGTCCCAATCCTTATA
1601  ---------+---------+---------+---------+---------+---------+---------+---------+---------+---------+  1700

      CAACCAACTTCCACAGGAAGGTCAGGTCGGGGACAACAAAAAAACAGGCAAGGGAAATTTTTTAATTTGGGTTGTCTTGTTTGCTGCATAATTTATGCAG
1701  ---------+---------+---------+---------+---------+---------+---------+---------+---------+---------+  1800

                                                                                         Poly A type sequences
                                                                                          ------------
      TAAAACACTACACATAACCCTTTTAGCAGTAGAGCAATGGTTGACCGTGTGCTTAGCTTCTTTTATTTTATTTTTTTATCAGCAAAGAATAAATAAAATA
1801  ---------+---------+---------+---------+---------+---------+---------+---------+---------+---------+  1900
                                                                                 HindIII
      ---                                                                        ------
      AAATGAGACACTTCAGGGATGTTTCAACCCTTATACAAAACCCCAAAAACAAGTTTCCTAGCACCCTACCAACTAAGCTT
1901  ---------+---------+---------+---------+---------+---------+---------+---------+  1980
```

Fig. 18. Nucleotide sequence of BZ-1 insert cloned into the expression vector pUC18pvPSPTexp. DNA isolated from clone pUC18pvPSPTBZ-1 was sequenced using pUC and the CPCR-engineering oligonucleotide primers. The use of extension sequencing reactions and 1 m gels [48,52] allowed for readings extending beyond 800 bp. Thus, the complete insert was sequenced using these oligonucleotide primers. The location of important restriction enzyme sites (*Hin*d I) and phaseolin gene regulatory elements are indicated above the sequence line. The location of the transcription initiation site is denoted by the # sign. The amino acid sequence of the high sulfur-containing polypeptide is given below the counting line.

References

1. Allison R, Johnston RE, Dougherty WG (1986) The nucleotide sequence of the coding region of tobacco etch virus genomic RNA: evidence for the synthesis of a single polyprotein. Virology 154: 9–20.
2. Altenbach SB, Pearson KW, Leung FW, Sun SSM (1987) Cloning and sequence analysis of a cDNA encoding a Brazil nut protein exceptionally rich in methionine. Plant Mol Biol 8: 239–250.
3. Altenbach SB, Pearson KW, Meeker G, Saraci LC, Sun SSM (1989) Enhancement of the methionine content of seed proteins by the expression of a chimeric gene encoding a methionine-rich protein in transgenic plants. Plant Mol Biol 13: 513–522.
4. An G (1986) Development of plant promoter expression vectors and their use for analysis of differential activity of nopaline synthase promoter in transformed tobacco cells. Plant Physiol 81: 86–91.
5. Beaucage SL, Caruthers MH (1981) Deoxynucleoside phosphoramidites – A new class of key intermediates for deoxypolynucleotide synthesis. Tetrahedron Lett 22: 1859–1862.
6. Benfey PN, Chua N-H (1989) Regulated genes in transgenic plants. Science 244: 174–181.
7. Benfey PN, Ren L, Chua N-H (1989) The CaMV 35S enhancer contains at least two domains which can confer different developmental and tissue-specific expression patterns. EMBO J 8: 2195–2202.
8. Carrington JC, Dougherty WG (1987) Processing of the tobacco etch virus 49 K protease requires autoproteolysis. Virology 160: 355–362.
9. Carrington JC, Cary SM, Parks TD, Dougherty WG (1989) A second proteinase encoded by a plant potyvirus genome. EMBO J 8: 365–370.
10. Chee PP, Klassy RC, Slightom JL (1986) Expression of a bean storage protein 'phaseolin minigene' in foreign plant tissues. Gene 41: 47–57.
11. Chen Z-L, Schuler MA, Beachy R (1986) Functional analysis of regulatory elements in a plant embryo-specific gene. Proc Natl Acad Sci USA 83: 8560–8564.
12. Domier LL, Franklin KM, Shahabuddin M, Hellmann GM, Overmeyer JH, Hiremath ST, Siaw MFE, Lomonossoff GP, Shaw JG, Rhoads RE (1986) The nucleotide sequence of tobacco vein mottling virus RNA. Nucl Acids Res 14: 5417–5430.
13. Dougherty WG, Carrington JC (1988) Expression and function of potyviral gene products. Ann Rev Phytopath 26: 123–143.
14. Dougherty WG, Cary SM, Parks TD (1989) Molecular genetic analysis of a plant virus polyprotein cleavage site: a model. Virology 171: 356–364.
15. Doyle JJ, Schuler MA, Godette WD, Zenger V, Beachy RN, Slightom JL (1986) The glycosylated seed storage proteins of *Glycine max* and *Phaseolus vulgaris*: structural homologies of genes and proteins. J Biol Chem 261: 9228–9238.
16. Erlich HA (1989) PCR technology: Principles and Applications for DNA Amplification. Stockton Press, New York.
17. Gallie DR, Sleat DE, Watts JW, Turner PC, Wilson TMA (1987) A comparison of eukaryotic viral 5′-leader sequences as enhancers of mRNA expression *in vivo*. Nucl Acids Res 15: 8693–8711.
18. Guilley H, Dudley RK, Jonard G, Balazs E, Richards KE (1982) Transcription of cauliflower mosaic virus DNA: detection of promoter sequences, and characterization of transcripts. Cell 30: 763–773.
19. Hatzenbuhler NT, Rehberg EF, Carter JB, Palermo DP, Hubert EV, Bergum PW, Erickson LA, Marotti KR (1989) Design and chemoenzymatic synthesis of a tissue plasminogen activator gene with unique restriction sites: a model for studying protein domain function. Nucleosides Nucleotides 8: 991–994.

20. Hepburn AG, White J, Pearson L, Maunders MJ, Clarke LE, Prescott AG, Blundy KS (1985) The use of pNJ5000 as an intermediate vector for the genetic manipulation of *Agrobacterium* Ti-plasmids. J Gen Microbiol 131: 2961–2969.
21. Innis MA, Gelfand DH, Sninsky JJ, White TJ (1989) PCR Protocols: A Guide to Methods and Applications. Academic Press, New York.
22. Jobling SA, Gehrke L (1987) Enhanced translation of chimaeric messenger RNAs containing a plant viral untranslated leader sequence. Nature 325: 622–625.
23. Kaufman DL, Evans GA (1990) Restriction endonuclease cleavage at the termini of PCR products. Biotechniques 9: 304–306.
24. Kay R, McPherson J (1987) Hybrid pUC vectors for addition of new restriction enzyme sites to the ends of DNA fragments. Nucl Acids Res 15: 2778.
25. Keohavong P, Thilly WG (1989) Fidelity of DNA polymerases in DNA amplification. Proc Natl Acad Sci USA 86: 9253–9257.
26. Kozak M (1986) Point mutations define a sequence flanking the AUG initiator codon that modulates translation by eukaryotic ribosomes. Cell 44: 283–292.
27. Lutcke HA, Chow KC, Mickel FS, Moss KA, Kern HF, Scheele GA (1987) Selection of AUG initiation codons differs in plants and animals. EMBO J 6: 43–48.
28. MacFerrin KD, Terranoma MP, Schreiber SL, Verdine GL (1990) Overproduction and dissection of proteins by the expression-cassette polymerase chain reaction. Proc Natl Acad Sci USA 87: 1937–1941.
29. Maiss E, Timpe U, Briske A, Jelkmann W, Casper R, Himmler G, Mattanovich D, Katinger HWD (1989) The complete nucleotide sequence of plum pox virus RNA. J Gen Virology 70: 513–524.
30. Maniatis T, Fritsch E, Sambrook J (1982) *Molecular Cloning*: A Laboratory Manual. Cold Spring Harbor Laboratory, Cold Spring Harbor NY.
31. Maxam AM, Gilbert W (1980) Sequencing end-labeled DNA with base-specific chemical cleavages. Meth Enzymol 65: 499–560.
32. Morelle G (1989) A plasmid extraction procedure on a miniprep scale. Focus 11: 7–8.
33. Mullis KB, Faloona F, Scharf SJ, Saiki RK, Horn GT, Erlich HA (1986) Specific enzymatic amplification of DNA *in vitro*: the polymerase chain reaction. Cold Spring Harbor Symp Quant Biol 51: 263–273.
34. Murray MG, Thompson WF (1980) Rapid isolation of high molecular weight plant DNA. Nucl Acids Res 8: 4321–4325.
35. Namba S, Ling K-S, Gonsalves C, Slightom JL, Gonsalves D (1991) Transgenic plants that express watermelon mosaic virus II, zucchini yellow mosaic virus, or papaya ringspot virus coat protein gene show different level of protection when challenge inoculated with various potyviruses (in preparation).
36. Ow DW, Jacobs JD, Howell SH (1987) Functional regions of the cauliflower mosaic virus 35S RNA promoter determined by use of the firefly luciferase gene as a reporter of promoter activity. Proc Natl Acad Sci USA 84: 4870–4874.
37. Pietrzak M, Shillito RD, Hohn T, Potrykus I (1986) Expression in plants of two bacterial antibiotic resistance genes after protoplast transformation with a new plant expression vector. Nucl Acids Res 14: 5857–5868.
38. Quemada H, Kearney C, Gonsalves D, Slightom JL (1989) Nucleotide sequences of the coat protein genes and flanking regions of cucumber mosaic virus strains C and WL RNA 3. J Gen Virol 70: 1065–1073.
39. Quemada H, L'Hostis B, Gonsalves D, Reardon I, Heinrickson R, Hiebert EL, Sieu LC, Slightom JL (1990) The nucleotide sequences of the 3′ terminal regions of papaya ringspot virus strains W and P. J Gen Virol 71: 203–210.
40. Quemada H, Sieu LC, Siemieniak DR, Gonsalves D, Slightom JL (1990) Watermelon mosaic

virus II and Zucchini yellow mosaic virus: cloning of 3′-terminal regions, nucleotide sequences, and phylogenic comparisons. J Gen Virol 71: 1451–1460.
41. Quemada H, Gonsalves D, Slightom JL (1991) Expression of cucumber mosaic virus strain-C coat protein gene: analysis of protection against infections by CMV strains -C, -WL, or -Chi using mechanical and aphid transmission vectors. Phytopathology, in press.
42. Rehberg EF, Theriault NY, Carter JB, Palermo DP, Hubert EV, Bergum PW, Wunderlich CJ, Erickson LA, Marotti KR (1989) Construction of a tissue plasminogen activator gene having unique restriction sites to facilitate domain manipulation. Protein Enging 2: 371–377.
43. Robaglia C, Durand-Tardif M, Tronchet M, Boudazin G, Astier-Manifacier S, Casse-Delbart F (1989) Nucleotide sequence of potato virus Y (N strain) genomic RNA. J Gen Virol 70: 935–947.
44. Rogers SO, Bendich A (1988) Extraction of DNA from plant tissues. In: Gelvin SB *et al.* (eds), Plant Molecular Biology Manual, Chap. A6. Kluwer Academic Publishers, Dordrecht, Netherlands.
45. Saghai-Maroof MA, Soliman KM, Jorgensen RA, Allard RW (1984) Ribosomal DNA spacer-length polymorphisms in barley: Mendelian inheritance, chromosomal location, and population dynamics. Proc Natl Acad Sci USA 81: 8014–8018.
46. Saiki RK, Gelfand DH, Stoffel S, Scharf S, Higuchi R, Horn GT, Mullis KB, Erlich HA (1988) Primer-directed enzymatic amplification of DNA with a thermostable DNA polymerase. Science 239: 487–491.
47. Scharf SJ, Horn GT, Erlich HA (1986) Direct cloning and sequence analysis of enzymatically amplified genomic sequences. Science 233: 1076–1078.
48. Siemieniak DR, Sieu LC, Slightom JL (1991) Strategy and methods for directly sequencing cosmid clones. Anal Biochem 192: 441–448.
49. Sinha ND, Biernat J, McManus J, Koster H (1984) Polymer support oligonucleotide synthesis XVIII: use of β-cyanoethyl-N, N-dialkylamino-/N-morpholino phosphoramidite of deoxynucleosides for the synthesis of DNA fragments simplifying deprotection and isolation of the final product. Nucl Acids Res 12: 4539–4557.
50. Slightom JL, Sun SM, Hall TC (1983) Complete nucleotide sequence of a French bean storage protein gene: plaseolin. Proc Natl Acad Sci USA 80: 1897–1901.
51. Slightom JL, Drong RF, Klassy RC, Hoffman LM (1985) Nucleotide sequences from phaseolin cDNA clones: the major storage proteins form *Phaseolus vulgaris* are encoded by two unique gene families. Nucl Acids Res 13: 6483–6498.
52. Slightom JL, Siemieniak DR, Sieu LC (1991) DNA sequencing: strategy and methods to directly sequence large DNA molecules. In: Miyamoto MM, Cracraft J (eds), Phylogenetic Analysis of DNA. Oxford University Press, Oxford, in press.
53. Slightom JL (1991) Custom polymerase-chain-reaction engineering of a plant expression vector. Gene 100, in press.
54. Tindall KR, Kunkel TA (1988) Fidelity of DNA synthesis by *Thermus aquaticus* DNA polymerase. Biochemistry 27: 6008–6013.
55. Yeh S-D, Gonsalves D (1985) Translation of papaya ringspot virus RNA in vitro: detection of a possible polyprotein that is processed for capsid protein, cylindrical-inclusion protein, and amorphous-inclusion protein. Virology 143: 260–271.
56. Yu MH, Frenkel MJ, McKern NM, Shukla DD, Strike PM, Ward CW (1989) Coat protein of potyviruses. 6. Amino acid sequence suggest watermelon mosaic virus 2 and soybean mosaic virus-N are strains of the same potyvirus. Arch Virol 105: 55–64.

Plant Molecular Biology Manual **C3**: 1–28, 1991.

Using polymerase chain reaction to identify transgenic plants

PAULA P. CHEE, ROGER F. DRONG & JERRY L. SLIGHTOM
Molecular Biology–Unit 7242, The Upjohn Company, Kalamazoo, MI 49007, USA

Introduction

The ability to transfer genes into many plant species is well established. The more commonly used techniques include *Agrobacterium tumefaciens* T-DNA-mediated transfer [3, 15, 16], direct DNA transfer into protoplasts [34], and microprojectile bombardment [21]. Identification of transformed plant tissues can be accomplished by a number of techniques including the use of plant-expressible, selectable-marker and/or reporter genes [28]. The most commonly used genes for selection are those encoding neomycin phosphotransferase II (NPTII) [10] and hygromycin phosphotransferase (*hpt*) [39]. By far the most commonly used reporter genes are those that encode chloramphenicol acetyltransferase (CAT) [10] and β-glucuronidase (GUS) [18].

The use of these selectable-marker and reporter genes has been very successful for the selection and identification of transformed tissues. However, there are some situations where the use of a selectable marker gene is not effective or where a reporter gene is not informative, either because of its lack of expression or due to its incomplete transfer. In other cases, the presence of the selectable marker and reporter genes is informative for the identification of transformed plants, but additional information is needed to determine if other genes within the transferred DNA construction (e.g. genes that express a beneficial trait) are also present. The early confirmation of a target gene in transformed plants could be very useful because its expression may take place at a much later stage of plant development (e.g. expressed in the flower or seed). There is a need for a technique to determine the presence and structural integrity of transferred target genes, so that plants that do not contain the target gene (or intact target gene) can be discarded early, thus conserving resources for plants that have a putative functional target gene.

In the past, the determination of the structural integrity of a gene could only be approached by the use of the genomic blot hybridization technique [37], whereby total genomic DNA is isolated from the plant tissue(s), digested with one or more restriction enzyme(s), subjected to electrophoresis through a gel matrix, and transferred onto a filter membrane (nitrocellulose or nylon). After fixing the DNA onto the filter membrane it is hybridized with a specific ^{32}P-labeled nucleotide probe (the complete or partial fragment of the gene or its cDNA component) followed by exposure to film and an analysis of the autoradiograph to determine if the targeted gene is contained within the expected size restriction enzyme fragments. Information gained from the genomic blot hybridization technique is extremely valuable for the determination of the structural integrity of transferred target genes and to show that the gene is actually integrated within the plant genome; however, the technique is too laborious to be used for screening a large number of putatively transformed plants. A rapid procedure capable of obtaining structural information is needed. After the identification of plants that contain the target gene, genomic blots can be used to verify integration and to determine the number of gene copies present on a reduced number of plants (see Section 3.3).

Within the last few years a new technique has been developed that allows the *in vitro* synthesis and amplification of specific segments of DNA. This technique, known as polymerase chain reaction (PCR) [26,32], can enzymatically amplify a specific region of DNA that is flanked by two oligonucleotide primers that share identity to the opposite DNA strands. Amplification is achieved by a repetitive series of cycles involving template denaturation, primer annealing, and DNA synthesis by DNA polymerase. By using repetitive cycles, where the primer extension products of the previous cycle serve as new templates for the following cycle, the number of target DNA copies has the potential to double each cycle. Thus if 20 cycles are performed, theoretically the target DNA could be amplified by a million-fold (2^{20}).

Originally, temperature control was achieved by transferring samples among three water baths set at the appropriate temperatures: denaturation (94 °C), primer annealing (37 to 65 °C), and DNA polymerase extension (37 to 40 °C). The polymerase used initially for DNA synthesis was the Klenow fragment of *Escherichia coli* DNA polymerase I; however, this enzyme is inactivated by the high temperature required for the denaturation of the template strands, and fresh enzyme had to be added during each cycle [32]. This requirement for the addition of new DNA polymerase was eliminated by using the thermostable DNA polymerase (*Taq*), isolated from *Thermus aquaticus* [33], which remains active even after being heated to temperatures near 94 °C. The use of *Taq* polymerase greatly simplified the PCR technique and, without the need to add new DNA polymerase after each cycle, it was only a short time before automated thermal cycling apparatuses became available to relieve one from standing by the water baths to make periodic transfers for what seemed to be an eternity (generally about 3 to 4 hours). However, it should be pointed out that one does not have to use an automated thermal cycling apparatus to perform PCR amplifications, because amplifications obtained by the manual transfer of samples among three water baths with temperature control is as precise as any of the automated thermal cycling apparatuses.

Unlike *E. coli* DNA polymerase I and T4 DNA polymerase, *Taq* polymerase does not appear to have a 3′ to 5′ exonuclease 'proof-reading' activity [14], which means that *Taq* polymerase does not have a similar fidelity for the synthesis of DNA. The error frequency for *Taq* polymerase was originally estimated by nucleotide sequence analysis of cloned PCR-generated DNAs to be very high, about 1 error in 400 nucleotides [33]. However, by obtaining the optimum reaction conditions the error rate has been found to be much less, about 1 in 4000 nucleotides [20]. Although the error rate is critical for PCR experiments that require a high degree of fidelity especially when used to engineer expression cassettes ([35], and see Chapter B16), it is not critical for the identification of genes transferred into transgenic plants.

Having the means to amplify a specific region of target DNA by a million-fold is indeed a very powerful method to determine the identity and structural integrity of a transferred target gene in transgenic plants. A unique property of the PCR technique is that only small amounts of target DNA (usually between 100 ng to 1 μg) are needed. The DNA need not be pure for amplification (provided the samples do not contain components that inhibit *Taq* polymerase) and the amplified DNA product can generally be detected by gel electrophoresis followed by staining with ethidium bromide (thus, in many cases radioactive probes are not needed). Because the sequence of the target gene and surrounding DNA is generally known, specific oligonucleotide primers can be synthesized to answer specific structural questions about the target gene.

However, PCR amplification is prone to generating its own false signals, the result of contaminants (from previous PCR amplifications) being present in the amplification reaction components (nucleotides, buffer, H_2O, etc.). The misinterpretation of positive results can also result from minute contaminations by any plasmid or λ clone that may contain target gene sequences. To avoid contamination problems careful laboratory procedures must be used: pre-aliquoting reagents (before any DNA is aliquoted), use of positive displacement pipettes, and physical separation of the area used for reaction preparation from the area where PCR amplifications are done. Contaminated reactions and reagents can generally be detected by including a series of 'blank' reactions; one that contains no template DNA and another that contains no primers. If contaminants become a major problem and result in false identification of transgenic plants, more drastic measures may be needed, such as performing all pipetting and vortexing in a hood [22]. Additional information concerning the use of PCR and its trouble shooting can be found in manuals dedicated specifically to the technique [13, 17].

The purpose of this chapter is to describe a procedure for the isolation of total plant DNA that is relatively simple, yet yields DNA pure enough to be used for PCR amplifications and restriction enzyme digestions. This procedure also yields sufficient quantities of DNA to permit the analysis of the target gene by the more laborious blot hybridization technique (see Section 3). The size of the preparation we describe generally yields between 100 to 200 μg of DNA, much more DNA than is necessary for PCR detection of the target gene. If target gene detection is all that is required, this DNA isolation procedure can be scaled down by a factor of 10 or more. We recommend that at least 2 μg of DNA be isolated in case several PCR amplifications are needed. We use the procedure at this large scale because once a particular transformed plant is identified as containing the target gene, the determination if it is integrated within the plant genome and its copy number by the genomic Southern blot technique can proceed rapidly (see Section 3).

Although the PCR technique can be used to obtain information concerning the organization of the target gene in the plant genome, this information is generally restricted to neighboring genes within the vector, which may have been integrated by the same event, such as neighboring selectable and/or reporter genes [3]. It would be a difficult task to use the PCR technique to show integration within a plant genome because the identity of the neighboring plant DNA, 5′- or 3′-flanking the target gene, is generally not known. Techniques such as inverse PCR [13, 17] can be used to amplify the target gene and flanking plant DNA; however, proving the identity of the flanking DNA ultimately requires the use of the genomic blot technique. In addition, in many cases the target gene could be imbedded within a large DNA structure (T-DNA region, vector DNA, etc.) and thus be located at a considerable distance from the flanking plant DNA. Techniques for PCR amplification of DNA fragments greater than 10 kb, let alone the more difficult task of inverse PCR, are still rare [13, 17]. We presently recommend the use of genomic blot hybridization technique to demonstrate integration of a target gene within a plant genome and to estimate the number of integrated gene copies.

Because of the reasons stated above, this chapter first describes a rapid DNA isolation procedure, followed by the use of PCR amplification to detect the presence of a target gene in transformed plants, and the use of genomic blot hybridizations to confirm the presence of the target gene and also to determine the number of copies transferred into the genome of the transformed plant.

Section 1. Plant DNA isolation

This plant DNA isolation procedure takes advantage of the method described by Murray and Thompson [27]. A solution of 1% hexadecyltrimethylammonium bromide (CTAB) is used to obtain DNA that is relative free of their major co-purifying carbohydrate contaminants (starches, pectins, charged polysaccharides, etc.). This original procedure was subsequently modified by Saghai-Maroof *et al.* [31] to permit the rapid isolation of DNA while retaining a relatively high degree of purity without the need for CsCl-ethidium bromide centrifugation. We have used this procedure to isolate DNA from leaf tissues of putatively transformed soybean [5], cucumber [6, 8, 36] tobacco [7], *Phaseolus vulgaris* [25] and squash (P.P. Chee, manuscript in preparation) plants. This procedure therefore appears to be well suited for a variety of plant species and tissue sources.

1.1. DNA isolation

Steps in the procedure

1. Collect leaves and wrap in cheese cloth, mark with appropriate identification numbers, then freeze in liquid nitrogen.
2. Transfer frozen tissue bundles into a freeze-dryer container and then connect to the freeze-dryer before leaves begin to thaw. Freeze-dry until the container reaches ambient room temperature (if cool to the touch, liquid is still present). This freeze-drying step can usually be done overnight depending on the strength of the vacuum source.
3. Grind tissues to a fine powder by using a Thomas-Wiley mill. After each sample has been ground, clean the mill so that the next sample will not be contaminated by residues from the previous samples (non-transformed and other special control tissues should be ground first).
4. Aliquot 300–400 mg fine powder into a 15 ml polypropylene centrifuge tube. Distribute tissue powder along sides of tube to avoid a clump of dry tissue in the bottom, and add 9.0 ml CTAB extraction buffer. Mix several times by inversion and vortex briefly. (If the amount of tissue is smaller, use appropriately reduced volumes of CTAB extraction buffer and other components listed throughout the procedure).
5. Incubate for 90 min in a 65 °C water bath; mix samples every 30 min by inversion several times.
6. Remove the tubes from the bath, wait 4–5 min to allow the tube to cool,

and then add 4.5 ml chloroform/octanol (24 : 1). Mix gently by inversion several times.

7. Centrifuge at room temperature in a table-top centrifuge for 10 min at full speed (3000 to 5000 rpm).
8. Transfer the top aqueous layer into new 15 ml polypropylene centrifuge tubes, add 4.5 ml chloroform/octanol (24 : 1), mix gently, and repeat step 7.
9. Pipette the top aqueous layer into 15 ml polypropylene tubes containing 6.0 ml cold isopropanol (2-propanol). Mix *very* gently by inverting.
10. After 10 to 15 min, remove precipitated DNA with a glass hook and transfer to 5 ml polypropylene tubes containing 3 ml of 76% EtOH, 0.2 M NaOAc. Leave the DNA on the hook in tubes for about 20 min.
11. Rinse the DNA on the hook briefly in 1–2 ml of 76% EtOH, 10 mM NH_4OAc and immediately transfer the DNA to 5 ml polypropylene tubes containing 1.0 ml TE (pH 8.0).
12. After the DNA has been dissolved, determine the DNA concentration by the use of TKO 100 Mini-Fluorometer (Hoefer Scientific Instruments).

Notes

2. Care must be taken in connecting the containers to the freeze-dryer because of the expanding nitrogen gas. We have obtained excellent freeze-drying results using 0.5-liter thick-wall Virtis containers and a Virtis Freeze Dryer Model 12.
3. We generally use the Thomas-Wiley mill in a chemical fume hood, which allows ease of cleaning by using an air hose to remove the remaining plant tissue powder. This cleaning method should only be done in a hood because the finely ground plant tissue will contaminate the laboratory.
3. If a Thomas-Wiley mill is not available the plant tissues can be ground using a pestle and mortar and ground glass. However, for a large number of samples this routine is quite tedious and thus a smaller sample size should be used (20 to 100 mg of tissue).
12. This size of DNA preparation generally yields between 100 to 200 μg of DNA.

Solutions

- CTAB extraction buffer (stock 1 liter)

– H_2O	730 ml (sterile, double-distilled)
– 1 M Tris-HCl pH 7.5	100 ml
– 5 M NaCl	140 ml
– 0.5 M EDTA pH 8.0	20 ml
– CTAB[1]	10 g
– β-mercaptoethanol[2]	10 ml

- 76% EtOH, 0.2 M NaOAc (stock 500 ml)

- Absolute EtOH 380 ml
  - 2.5 M NaOAc pH 6.3 40 ml
  - H_2O 80 ml (sterile, double-distilled)
- 76% EtOH, 10 mM NH_4OAc (stock 100 ml)
  - Absolute EtOH 76 ml
  - 1 M NH_4OAc 1 ml
  - H_2O 23 ml (sterile, double-distilled)
- TE pH 8.0 buffer (stock 100 ml)
  - 1 M Tris-HCl pH 8.0 1.0 ml
  - 0.5 M EDTA pH 8.0 0.2 ml
  - H_2O 98.8 ml (sterile, double-distilled).

[1] CTAB is hexadecylcetyltrimethylammonium bromide (Sigma H-5882); add to solution and stir at warm temperature until completely dissolved.

[2] Add β-mercaptoethanol to buffer just prior to use; 1% v/v = 140 m*M*.

Section 2. Identification of a target gene by PCR amplification

In this section we describe the use of PCR amplification to identify the presence of a target gene in putatively transformed tissues. We have used PCR amplifications to identify the presence of several specific genes: in soybean (the Nos-NPTII gene [5]), in cucumber (the cucumber mosaic virus and papaya ringspot virus coat protein genes; P.P. Chee, unpublished), and squash (the cucumber mosaic virus coat protein gene; P.P. Chee, unpublished). In this example we describe the use of two specific oligonucleotide primers to amplify a specific region of a target gene that is well known in the literature, the chimeric plant-expressible Nos-NPTII gene constructed by An *et al.* [2].

This chimeric NPTII gene uses the *Agrobacterium* T-DNA-derived nopaline gene (Nos) regulatory regions (promoter and polyadenylation signal) to direct the expression of the *E. coli* transposon Tn*5*-derived NPTII gene [2]. The oligomer primers used (21 and 24 nucleotides in length) were selected to amplify by PCR a DNA product of a size that would indicate that the transferred Nos-NPTII gene was intact (containing the Nos promoter and complete NPTII gene coding region). The 5′ primer, referred to as primer A (5′-CCCCTCGGT-ATCCAATTAGAG-3′), shares identity with the Nos promoter region 33 bp 5′ of the Nos gene translation initiation codon (ATG) [11]. The 3′ primer, referred to as primer B (5′-CGGGGGGTGGGCGAAGAACTCCAG-3′), shares identity with the opposite strand of 3′-flanking region of the NPTII gene, 150 bp 3′ of the NPTII translation termination codon (TAA) [24]. The DNA product amplified by these primers is about 900 bp in length; this was confirmed by PCR amplification using 10 ng of pGA482 DNA as a positive control template (Fig. 1A).

Detection of the Nos-NPTII gene in DNAs isolated from transgenic plant (in this case, soybean) was done by using approximately 200 ng of total soybean DNA [5]. PCR amplification cycles were controlled by a Perkin-Elmer-Cetus DNA Thermal Cycler. PCR samples were subjected to 30 amplification cycles and the presence or absence of the characteristic DNA fragment was determined by loading 1/2 of each reaction sample onto an agarose gel (0.7%) followed by gel electrophoresis and ethidium bromide staining. If necessary, confirmation of the identity and the detection of the characteristic DNA band in weakly amplified samples can be accomplished by subsequent transfer of these DNAs onto nylon filters and hybridization with a ^{32}P-labeled [23] NPTII DNA fragment [5]. We have found the 600 bp fragment generated by a *Bgl* II and *Nco* I digest [23] to be excellent radioactive probe fragment because of its low background when hybridized against PCR-generated products or total plant DNAs ([5]; see Section 3).

Many different buffer and nucleotide concentration conditions have been used for amplifying DNA by PCR [13, 17], and some of these differences may be necessary for successful DNA amplification. However, we have not found that the amplification of target DNAs within the plant species described above require any special buffer or nucleotide concentration modifications. We have been satisfied with the buffer and nucleotide concentrations recommended by Perkin-Elmer-Cetus (see Solutions).

2.1. PCR amplification of a target gene

Steps in the procedure

1. From the DNA concentration of each sample, determine the volume needed to obtain 200 ng; also from the concentration of each PCR primer (concentrations are determined by measuring OD_{260}, and 1 OD = 20 μg/ml [23]) determine the volume needed to obtain a final concentration of between 0.5 to 1 μM in a total reaction volume of 100 μl.
2. For each PCR sample use a sterile 0.5 ml centrifuge tube, label each tube accordingly. The order in which the reaction components are added is important; generally add the component with the largest volume first (usually H_2O), followed by the buffer and dNTP mix, then add the DNAs (genomic and oligonucleotide primers), and add the *Taq* polymerase last.
3. Assuming the genomic DNA concentration is 200 μg/ml and the concentration of each oligomer primer is 200 μg/ml, the following components are added to each PCR sample tube:

Component	*Order of addition*	*Volume (μl)*
Double-distilled sterile H_2O	1	66.5
PCR Buffer (10×)	2	10.0
dNTP Mix, 1.25 mM (each dNTP)	3	16.0
Primer A (200 μg/ml)	4	3.0
Primer B (200 μg/ml)	5	3.0
Genomic DNA (200 μg/ml)	6	1.0
Taq polymerase (2.5 units/μl)	7	0.5
	Total v	100.0

4. Mix gently, then briefly centrifuge. Overlay each PCR sample with 100 μl of mineral oil to prevent evaporation.

5. Ensure that the temperature controls are in place, i.e., if water baths are used, they should be at the appropriate temperatures; if an automated temperature cycler is used, the appropriate program should already be selected. Recommended temperatures and times used for PCR amplification of the Nos-NPTII gene are from Chee *et al.* [5]. The following Perkin-Elmer-Cetus DNA Thermal Cycler program was used:

Duplex denaturation	94 °C	2 min
Annealing	55 °C	2 min
DNA synthesis	72 °C	3 min.

6. Repeat the temperature cycles for the appropriate number of times; in the case of PCR amplification of the Nos-NPTII gene, 30 cycles were used. On the last cycle omit the heat denaturation step and extend the DNA synthesis step to 10 min to aid strand completion of the accumulated DNAs. At this point the samples can be analyzed or, if desired, the samples can be stored refrigerated or frozen at −20 °C. If a refrigerated automated thermal cycler is used it can be programmed to maintain the samples at 4 °C overnight for analysis the following day.
7. Analysis of the PCR reaction products is very straightforward depending on the size of the expected DNA product. About 1/2 (50 μl) of each sample is subjected to electrophoresis through either a 0.7% agarose gel (if the DNA product is larger than 700 bp) or a 7% acrylamide-15% glycerol gel (if the DNA product is smaller than 700 bp). After electrophoresis, the gels are stained with ethidium bromide (a solution containing about 0.5 ng/ml), observed by placing on a UV light box, and photographed (Fig. 1).

Notes

1. For oligonucleotide primers in the range of 20 nucleotides, use between 300 to 600 ng per reaction. If the final concentration of the oligonucleotide is to be about 1 μM, then the number of moles of the oligonucleotide is calculated as follows: 1×10^{-6} mol/l $\times$ 1×10^{-4} l $= 1 \times 10^{-10}$ mol. Assuming the average molecular weight of a nucleotide of 300 g/mol, the molecular weight of a oligonucleotide of 20 nucleotides is about 6000 g/mol. Thus, the mass of the oligonucleotide to be added is 1×10^{-10} mol $\times$ 6000 g/mol $=$ 600 ng.
7. PCR products from plants containing the target gene should yield a visible DNA band at the expected size, as shown in Fig. 1 for the amplification of the Nos-NPTII gene transferred into the soybean genome [5]. The lack of a visible band should not be taken that the sample does not contain the target gene, as it could be an indication of a poor PCR amplification due to having impure genomic DNA, or having made an error in measuring its concentration. Before discarding the sample (and putative transgenic plant) the gel containing the sample should be analysed by hybridization with ^{32}P-labeled target DNA, in this case, a DNA fragment from the NPTII gene. Hybridization results should determine if there is any small

amount of PCR product present and also confirm the identity of the band in samples with visible DNA bands. Before discarding the sample, PCR amplifications should be repeated in an experiment in which the amount of genomic DNA is increased (after rechecking that its concentration was correctly measured).

Solutions

- dNTP mix (Perkin-Elmer-Cetus)

– dATP 10 mM	125 μl	
– dGTP 10 mM	125 μl	
– dCTG 10 mM	125 μl	
– dTTP 10 mM	125 μl	
– H_2O	500 μl	(sterile, double-distilled)
– Total	1.0 ml	dNTP mix, 1.25 mM final concentration for each dNTP.

- PCR Buffer 10× (Perkin-Elmer-Cetus)

– KCl	500 mM
– Tris-HCl (pH 8.3)	100 mM
– $MgCl_2$	15 mM
– Gelatin	0.1% (w/v)

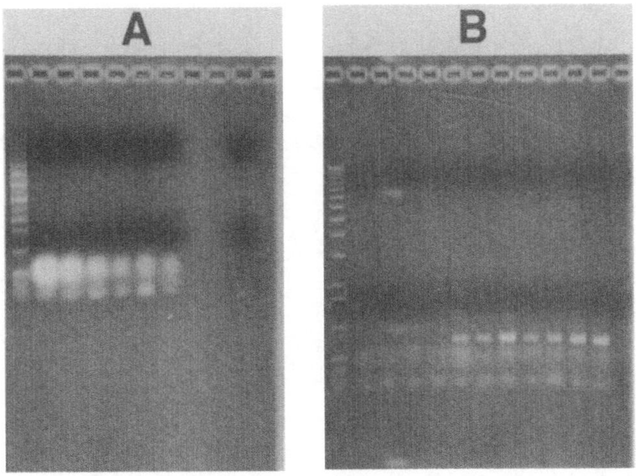

Fig. 1. PCR amplification of the Nos-NPTII gene contained within the pGA482 plasmid control DNA and the analysis of R_1 putative transformed soybean plants containing the T-DNA region of pGA482. Frame A: Analysis of PCR amplification of Nos-NPTII gene in pGA482 and a check of solutions for the presence of contaminants. From left to right, lane 1 contains the 1 kb DNA ladder size standard from Bethesda Research Laboratories. Lanes 2 to 7 are the results of loading 1/2 of the PCR amplification samples using various amounts of pGA482 as starter template (20, 10, 5, 3, 2, 1 pg, respectively). The expected DNA amplification product was visualized after staining with ethidium bromide. Lane 10 contains the PCR amplification control in which no template DNA was added to ensure that no pGA482 DNA is present in the PCR reagents (see list of solutions). Lanes 8, 9, and 11 were intentionally left blank. Frame B: PCR analysis of R_1 soybean plants derived from pGA482-transformed soybean obtained by Chee *et al.* [5]. This analysis was used to follow the segregation of the transformed phenotype in the R_1 seed population. From left to right, lane 1 contains the 1 kb DNA size standard and lanes 2 and 3 show the PCR amplification results of control (non-transformed) soybean DNA, 200 μg, which was used to show that the PCR reagents used were not contaminated. Lanes 4 to 12 show PCR amplification results for putative transformed soybean progeny from independent PCR analysis. Lane 13 contains 50 ng of pGA482 which has been digested with *Hind* III and *Bam* HI. This sample provides a positive control for checking the identity of the PCR-amplified products by hybridization to ^{32}P-labeled Nos-NPTII gene (see [5]).

Section 3. Genomic blot hybridization analysis of a target gene

The identification of transgenic plants by using PCR amplification of the target gene is an important step because it eliminates the need to carry forward a large number of putatively transformed plants, thus saving space and labor. The primary question concerning a newly transferred target gene is whether or not it is functional. The answer to this question depends on the stage of plant development in which the target gene is expressed; if the gene is expressed in root, stem, or leaf tissues, the answer can be obtained early. However, if the target gene is expressed later in plant development (flower, seed pod, or seed) or if the gene is to be expressed as the result of a physical or biological challenge, the answer to this question may be delayed for quite some time. In the meantime more information can be learned about the nature of the target gene in its new environment, and this information could be important in predicting whether the gene will be functional. Although the PCR analysis described above indicates that the target gene is intact, knowledge of its flanking DNA could be used to determine if the transfer was made as expected, i.e., T-DNA transfers would be expected to proceed from the right border (T_R) to the left border (T_L) [40], or if rearrangements have occurred. In addition, knowledge of the number of copies of the target gene in a particular transgenic plant line is important so that the level of expression obtained from each transferred target gene can be determined. Multiple gene transfers have been found using the *Agrobacterium*-mediated gene transfer mechanism [19], and DNA transfers made by electroporation [29] and microprojectile bombardment [9] quite frequently yield long tandem arrays of inter-dispersed gene and plasmid vector DNAs.

The genomic blot hybridization technique has previously been described by Dunsmuir *et al.* [12], but their description did not include DNA transfer by electroblotting or the use of the more durable nylon filters. Section 3.1 and 3.2 describe electroblotting of genomic DNA onto nylon filters and the hybridization of these filters with ^{32}P-labeled DNA probes. In Section 3.3 we expand on the determination of gene copy number by presenting an example for the Nos-NPTII gene transferred into the tobacco genome.

3.1. Electroblotting genomic DNA onto nylon filters

Steps in the procedure

1. After electrophoresis has been completed, stain the DNA with ethidium bromide and photograph the gel (Fig. 2). Place a ruler alongside the gel

so that the distance that any given band of DNA has migrated can be read directly from the photographic image.

2. Transfer the gel to a pyrex dish or suitable plastic box and trim away any unused areas of the gel with a razor blade.
3. Denature the DNA by immersing the gel in several volumes of 0.25 M NaOH for 20 min at room temperature with constant agitation.
4. Discard the solution and rinse the gel briefly in deionized water, then soak in 0.5× TBE buffer for 20 min at room temperature with constant agitation.
5. Fill the electroblot tank (Hoefer Model Transphor unit TE 52 or 42) with about 4 liters of 0.5× TBE.
6. Open the cassette by sliding the two halves past each other until the hook on each side disengages from the other half. Place one half in a big tray such that the hook faces up.
7. Place one wet porous pad (e.g., Scotch-brite) on the cassette half. Press on the pad to force out any trapped bubbles.
8. Lay two pieces of blot paper pre-wet with 0.5× TBE on the sponge.
9. Lay the gel on the blot paper. Make sure there are no air bubbles between the blot paper and the gel.
10. Cut a piece of nylon filter (Schleicher & Schuell Nytran, Amersham Hybond-N, Du Pont GeneScreen, or others) to the size of the gel. Wear gloves when handling the nylon filter. Use a ball-point pen (permanent ink) to draw a line on the filter to indicate the position of the gel wells; also write down your name, date and experiment number.
11. Float the filter in deionized water until completely wet, then soak it in 0.5× TBE until it is ready to use.
12. Place the wet nylon filter on top of the gel so that the line on the filter is on the line of slots at the top of the gel. Be careful to remove all air bubbles that are trapped between the gel and the filter.
13. Place one or two additional pieces of wet blot paper on top of the nylon filter. Again remove all air bubbles.
14. Place another wet porous pad on top of the blot paper without trapping air bubbles.
15. Place the second half of the cassette on top of the stack such that the hook is down and faces the hole near the edge of the bottom half.
16. Press the two halves together and slide them toward each other making sure both hooks engage with the opposite half. The assembled cassette can be secured tightly with pins, clamps or string.

17. Insert the assembled cassette into the TE unit with the membrane on the side of the gel facing the anode.
18. Add enough 0.5× TBE to cover the blotting assembly.
19. Place the TE power lid on top of the chamber for the TE 52 unit. The banana plugs on the electrode panels fit into the holes located in the underside of the power lid. The anode (+) is at the rear of the power lid.
20. If using the TE 42 unit with a separate power supply, place the lid on the chamber so that the red lead contacts the electrode that functions as the anode. Connect the leads to the power supply.
21. Plug in the power supply and turn on the power switch. Electroblot the DNA sample onto the nylon filter by using 20 V for 12 h at 4 °C. This temperature can be maintained by using a chilled H_2O source, or by using the elctroblotter in a cold room. Do not permit the buffer in the chamber to heat.
22. After electroblotting, pull the assembled cassette out, dismantle the blotting stack down to the filter, remove the filter and the gel together, place the filter side down on a clean pyrex dish or suitable plastic container.
23. Peel off the gel and re-stain it in ethidium bromide to determine if most of the DNA has been transferred (Fig. 2). Soak the filter in denaturation solution (1.5 M NaCl, 0.5 M NaOH) for 20 min with slow agitation.
24. Discard the solution and replace it with neutralization solution (1.5 M NaCl, 0.5 M Tris-Cl, pH 8.0). Leave the filter in neutralization solution for 20 min.
25. Place the wet filter with the DNA side up under a UV source such as a transilluminator for 3 min or an ultraviolet cross-linking device to cross-link the DNA to the filter.
26. The filter can be used immediately for hybridization. If not used immediately the filter can be wrapped with Saran Wrap and stored at 4 °C.

Notes

1. About 10 μg of total genomic DNA (haploid genome = 3×10^9 bp) must be applied to a single gel slot in order to be able to detect sequences that occur only at the single-copy level.
3. Depurination of the DNA sample by treatment in 0.25 M HCl is not necessary when DNAs are electroblotted.
21. The electroblot settings described are sufficient to transfer DNAs in the range of 20 kb. If larger DNA fragments are to be transferred, the voltage or the electroblotting time should be increased. However, if the voltage is increased, be sure that the apparatus is kept cool.

Solutions

- Denaturation solution
  - 0.5 M NaOH
  - 1.5 M NaCl
- Neutralization solution
  - 1.5 M NaCl
  - 0.5 M Tris-HCl (adjust pH to 7.0)
- 10× TBE electrophoresis buffer (4 liters)
  - Tris (4 M) 484.4 g
  - Boric acid (3.3 M) 205.4 g crystals
  - EDTA (5 mM) 14.9 g
  - Dissolve in a final volume of H_2O, check the pH, which should be about 8.3.

3.2. Genomic blot hybridization

Steps in the procedure

1. Place the filter in a flat-bottomed plastic container containing prehybridization solution (6× SSC, 1× Denhardt's reagent, 0.5% SDS), put the lid on the container and agitate at 68 °C for at least 2 h.
2. Denature the probe by adding 0.1 volume of 5 M NaOH and incubate at 65 °C for 15 min.
3. Add 0.1 volume of 5 M HCl followed by 0.1 volume of 2 M Tris-HCl (pH 7.5).
4. Remove the prehybridization solution from the container, add hybridization solution [6× SSC, 1× Denhardt's reagent, 2.0% SDS, 50 μg/ml polyadenylic acid (Sigma P-9403)], then add the probe at a concentration of 1×10^6 cpm/ml of hybridization solution for the detection of single-copy genes. More abundant DNA sequences can be detected at 5×10^5 cpm/ml.
5. Agiate gently at 68 °C for 12 h.
6. Remove the hybridization solution, rinse with 3× SSC, and wash the filter in 300 ml of 3× SSC for 1 h at 65 °C with constant gentle agitation.
7. Discard the solution and rewash the filter in 300 ml of 1× SSC for 1 h at 65 °C with constant gentle agitation.
8. Dry the filter at room temperature on a sheet of Whatman 3MM paper.
9. Wrap the filter in Saran Wrap and apply to X-ray film to obtain an autoradiographic image. Store the exposed film at −80°. The film exposure resulting from the genomic blot hybridization of the ^{32}P-labeled NPTII

gene against DNAs isolated from cucumber plants that were putatively transformed with the T-DNA region of the binary vector pGA482 is shown in Fig. 3.

10. After exposure, save the filters as they can be used many times, either for hybridizations with the same probe (if a higher specific activity probe is needed) or with a different gene probe.

Notes

9. Film exposures are generally done using two Quanta III intensifying screens (DuPont) and an exposure time ranging between 24 to 72 h.
10. Hybridized DNA probes can be removed from the filters by washing in distilled H_2O at 68 °C for about 30 min, or by a brief treatment with 0.2 M NaOH at room temperature.

Solutions

- 6× SSC
  - 0.9 M NaCl
  - 0.09 M sodium citrate (adjust pH to 7.2)
- 1× Denhardt's reagent
  - 0.02% BSA (bovine serum albumin; Sigma, A-4503)
  - 0.02% Ficoll (Sigma, F-4375)
  - 0.02% PVP (polyvinylpyrrolidone, Sigma, PVP-40)

3.3. Gene copy number determination by blot hybridization gene copy number reconstruction

Gene copy number determinations are generally made by using either solution hybridization kinetics (also referred to as C_OT. [4]) or by using genomic blot hybridizations. In both cases gene copy numbers are calculated using a gene reconstruction that takes into consideration the size of the species genome and the size of the plasmid vector (that contains the target gene), which is used to reconstruct the copy number equivalence. The accuracy of both methods is similar $\pm$ 0.5 to 1 gene copy [38]; however, the solution hybridization technique is complex and laborious. In contrast, genomic blot hybridization experiments are simpler and can be much more informative because the selection of different restriction enzymes can yield additional data concerning the DNA structures flanking the target gene. In this section we describe how to set-up a genomic blot copy number reconstruction and demonstrate its use to calculate the Nos-NPTII gene copy number contained within transgenic R_1 tobacco plants.

3.3.1. Determination of Nos-NPTII gene copy number in transformed R_1 tobacco plants

In this example we have transferred the T-DNA region of the *Agrobacterium* binary plasmid pGA482 [1] into tobacco (*Nicotiana tabacum*) tissues. Transformed R_0 tobacco plants were selected by growth on 500 μg kanamycin. These R_0 tobacco plants were advanced to produce R_1 plants which were analyzed by genomic blot hybridization to determine the copy number of the Nos-NPTII gene integrated with the genome.

Tobacco DNA was isolated using the procedure described in Section 1 (except that this DNA was centrifuged to equilibrium in an ethidium bromide-CsCl gradient), followed by digestion with two restriction enzymes, *Bam* HI and *Stu* I, which yield a 2.2 kb band [24] that can be detected using ^{32}P-labeled Nos-NPTII gene sequences. For this experiment the radioactive labeled probe was made by nick-translating [23] a 600 bp *Bgl* II to *Nco* I fragment of the NPTII gene [24]. Gene copy reconstruction of the Nos-NPTII was achieved by using the *Agrobacterium* binary plasmid, pGA482.

Steps in the procedure

1. Digest 10 μg of each purified tobacco DNA sample with *Bam* HI (66 units) and *Stu* I (33 units); digest at 37 °C for 2 h.
2. Calculate the amount of pGA482 DNA to add to the copy number reconstruction samples; this estimate is made first by determining the number of genomes contained within the genomic DNA digestion sample and by determining the mass of the reconstruction plasmid DNA needed to represent a number of molecules equivalent to the number of genomes present in the genomic DNA sample. For *Nicotiana tabacum*:

   3.9 pg [30] = 1 genome, thus 10 μg = 2.6×10^6 genomes

   For the reconstruction plasmid, pGA482, we need to calculate the equivalent number of molecules to represent the number of genomes in this range pGA482 contains about 1.63×10^4 bp and its mass is calculated by multiplying the number of bp with the molecular mass of each bp (700 Da per bp), i.e. $1.63 \times 10^4 \times 700$ Da = 11400 kDa per molecule. Thus: 11.4×10^6 g/mol = 6×10^{23} total molecules; 11.4 pg = 6×10^5 molecules; 43 pg = 2.6×10^6 molecules.
3. Set up a series of reconstruction *Eco* RI digestion samples using the plasmid pGA482 as follows:

Mass of plasmid pGA482	*Genome equivalence*
86 pg	2

172 pg	4
344 pg	8

4. After digestion of the genomic and reconstruction samples, subject them to electrophoresis through a 0.7% agarose gel. Transfer the DNA within the gel onto a nylon filter using the electroblotting procedure described in Section 3.1.
5. Hybridize the electroblotted filter with the appropriate DNA probe as described in Section 3.2. After hybridization and filter washing, expose the nylon filter to film. Figure 4 shows the hybridization results obtained for this experiment.
6. Determine the band intensities for hybridization bands that appear in the sample lanes that contain the genomic and reconstruction DNAs. This can be done using a densitometer, such as the LKB model 2202, or a Molecular Dynamics film scanner.
7. Using the integrated areas under the reconstruction hybridizations bands, determine the gene copy number for each hybridizing band found within the genomic DNA samples.

Notes

3. To minimize any processing errors, the plasmid DNAs used for the reconstruction should be added to 10 μg of genomic DNA isolated from untransformed tobacco tissues.
7. Gene copy number estimates were calculated using the Molecular Dynamics scan densities to construct a linear plot (least squares). The following data were obtained for the samples shown in Fig. 4:

Lane designation	*Integrated intensity*	*Gene copies*
2	950	2 (reconstruction)
4	1144	4 (reconstruction)
8	3169	8 (reconstruction)
A	122	0.25 (calculated)
B	879	2.2 (calculated)
C	237	0.5 (calculated)
D	1248	3.4 (calculated)
E	151	0.4 (calculated)
F	59	0.1 (calculated)

These results first show that most of these R_1 transformed tobacco plants are indeed transformed, containing the Nos-NPTII gene of pGA482. Secondly, they show that not all of the transformants received the same number of Nos-NPTII genes. Some of the plant lines appear to be the result of multiple T-DNA integrations [19], while others appear to have less than one complete gene. The latter observation may be due to poor DNA transfer or hybridization for specific samples, or could indicate that the plant is not homozygous for the transgenic trait, and is a mosaic in which the transgenic trait is segregating. Because of the mosaic nature of many of the plants derived from most transformation techniques, more accurate gene copy number estimates can be obtained from plants in the R_2 generation as these plants are more likely to be homozygous for the transgenic trait.

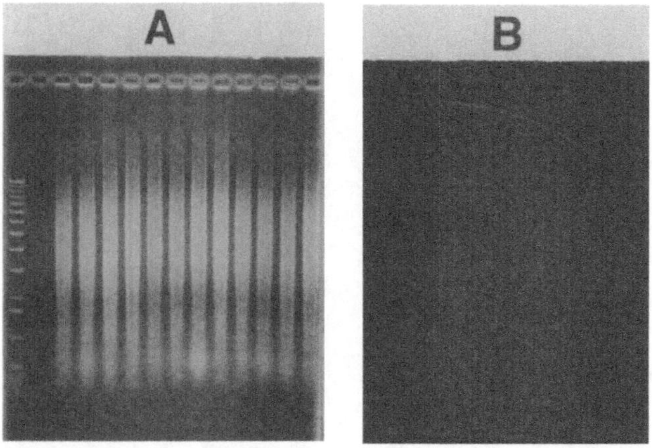

Fig. 2. Electroblotting of restriction enzyme digested DNA samples isolated from R_0 cucumbers that have been transformed with the binary vector pGA482. Frame A: Cucumber tissues were transformed with the disarmed *Agrobacterium* strain C58Z707 that contained the *Agrobacterium* binary plasmid pGA482 [1]. From left to right, lane 1 (labeled M in Fig. 3) contains the 1 kb ladder size standard and lane 2 contains 1 ng of pGA482 DNA that has been digested with *Hind* III and *Bam* HI. This control pGA482 DNA was used to help identify the correct size of the NPTII gene within the DNAs isolated from the transformed cucumber plants. Lanes 3 to 12 each contain 7 μg of cucumber DNA (each isolated from an individual R_0 plant) that has been digested with *Hind* III and *Bam* HI. DNAs were isolated from R_0 cucumber plants T45, T46, T48, T49, T52, T53, T54, T55, T56, and T58, respectively. Lanes 13 and 14 contain 7 μg of DNA isolated from non-transformed cucumber plants. The picture shows the ethidium bromide-stained pattern after electrophoresis through a 0.7% agarose gel. Frame B shows the ethidium bromide-stained pattern of this 0.7% agarose gel after electroblotting using the procedure described in Section 3.1.

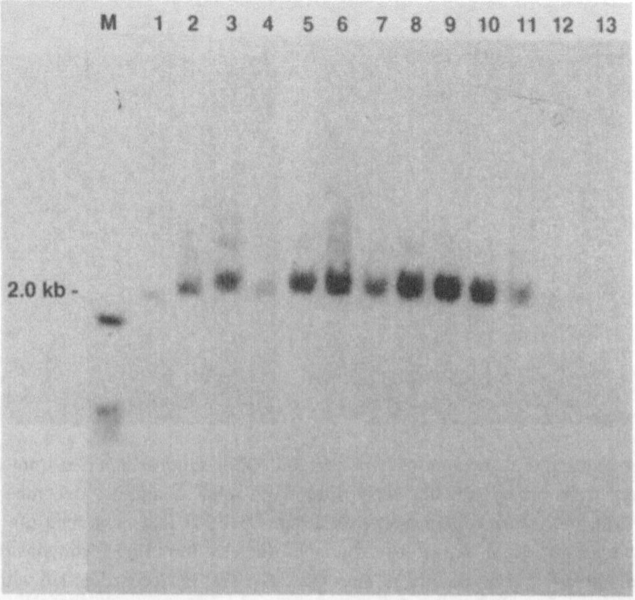

Fig. 3. Genomic blot hybridization analysis of DNAs isolated from R_0 cucumber plants. The nylon filter obtained after electroblotting the *Hind* III- and *Bam* HI-digested cucumber DNAs shown in Fig. 2. This nylon filter was hybridized with ^{32}P-labeled NPTII gene and autoradiographed as described in Section 3.2. The results show bands in most of the sample lanes. The bands found in the lane labeled M, which contains the 1 kb size standard, is due to the presence of *E. coli*-derived plasmid DNAs used in the DNA fragment mixture. The positive control pGA482-derived 2 kb *Hind* III-*Bam* HI hybridizing fragment is shown in lane 1. The lanes labeled 2 to 12 show the hybridization results for DNAs isolated from R_0 cucumber plants T45, T46, T48, T49, T52, T53, T54, T55, T56, and T58. Only one negative control is shown (lane 13), which contains DNA isolated from a non-transformed cucumber plant.

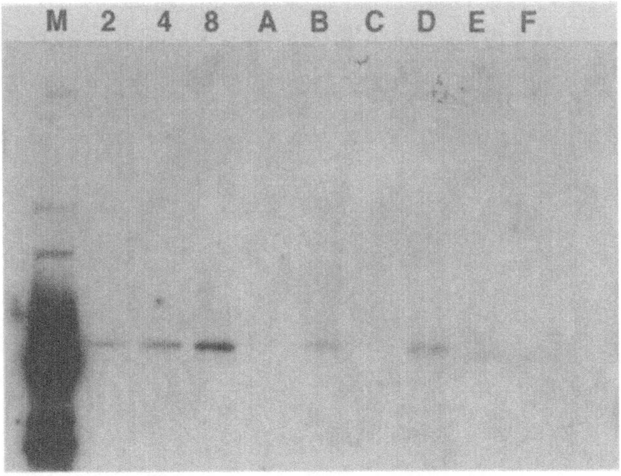

Fig. 4. Determination of gene copy number for the Nos-NPTII gene in tobacco plants transformed with the T-DNA region from pGA482. Ten μg of DNA (isolated from leaf samples as described in Section 1) was digested with 66 units of *Bam* HI and 33 units of *Stu* I and subjected to electrophoresis through a 0.7% agarose gel. DNA in this gel was transferred to a nylon filter and hybridized with the ^{32}P-labeled 600 bp *Bgl* II to *Nco* I fragment of the NPTII gene [24]. The lane labeled (M) contains the 1 kb size standard, and the lanes labeled 2, 4, and 8 represent the respective gene copy reconstructions and contain 86, 172, 344 pg of the plasmid pGA482. The lanes labeled A to F contain the DNAs isolated from the various R_1 tobacco plants. The hybridization was done as described in Section 3.2. Hybridization signals are not present in all the tobacco DNA plants analyzed because the transformed phenotype is heterozygous in these R_1 plants. Gene copy numbers were calculated after determining the hybridization intensity present in the reconstruction and sample lanes (by integrating the density of the scanning signal for the band present in each lane using a Molecular Dynamic scanner), see Section 3.3.1, step 7.

References

1. An G (1986) Development of plant promoter expression vectors and their use for analysis of differential activity of nopaline synthase promoter in transformed tobacco cells. Plant Physiol 81: 86–91.
2. An G, Watson BD, Stachel S, Gordon MP, Nester EW (1985) New cloning vehicles for transformation of higher plants. EMBO J 4: 277–288.
3. An G, Ebert PR, Mitra A, Ha SB (1988) Binary vectors. In: Gelvin SB *et al.* (eds) Plant Molecular Biology Manual, Chap. A3. Kluwer Academic Publishers, Dordrecht, Netherlands.
4. Britten RJ, Graham E, Neufeld BR (1974) Analysis of repeating DNA sequences by reassociation. Meth Enzymol 29E: 363–406.
5. Chee PP, Fober KA, Slightom JL (1989) Transformation of soybean (*Glycine max*) by infecting germinating seeds with *Agrobacterium tumefaciens.* Plant Physiol 91: 1212–1218.
6. Chee PP (1990) Transformation of *Cucumis sativus* tissue by *Agrobacterium tumefaciens* and the regeneration of transformed plants. Plant Cell Rep 9: 245–248.
7. Chee PP, Jones JM, Slightom JL (1991) Expression of bean storage protein minigene in tobacco seeds: introns are not required for seed specific expression. J. Physiol. Plant 137: 402–408.
8. Chee PP, Slightom JL (1991) Transfer of cucumber mosaic virus coat protein gene into the genome of *Cucumis sativus* and analysis of its expression. J Hort, in press.
9. Christou P, Swain WF, Yang N-S, McCabe DE (1989) Inheritance and expression of foreign genes in transgenic soybean plants. Proc Natl Acad Sci USA 86: 7500–7504.
10. De Block M, Herrera-Estrella L, Van Montagu M, Schell J, Zambryski P (1984) Expression of foreign genes in regenerated plants and in their progeny. EMBO J 3: 1681–1689.
11. Depicker A, Stachel S, Dhaese P, Zambryski P, Goodman HM (1982) Nopaline synthase: transcript mapping and DNA sequence. J Mol Appl Genet 1: 561–573.
12. Dunsmuir P, Bond D, Lee K, Gidoni D, Townsend J (1988) Stability of introduced genes and stability in expression. In: Gelvin SB *et al.* (eds) Plant Molecular Biology Manual, Chap. C1. Kluwer Academic Publishers, Dordrecht, Netherlands.
13. Erlich HA (1989) PCR Technology: Principles and Applications for DNA Amplification. Stockton Press, New York.
14. Gelfand DH (1989) *Taq* DNA polymerase. In: Erlich HA (ed) PCR Technology: Principles and Applications for DNA Amplification, pp. 7–22. Stockton Press, New York.
15. Hooykaas PJJ (1988) *Agrobacterium* molecular genetics. In: Gelvin SB *et al.* (eds) Plant Molecular Biology Manual, Chap. A4. Kluwer Academic Publishers, Dordrecht, Netherlands.
16. Horsch RB, Fry J, Hoffmann N, Neidermeyer J, Rogers SG, Fraley RT (1988) Leaf disc transformation. In: Gelvin SB *et al.* (eds) Plant Molecular Biology Manual, Chap. A5. Kluwer Academic Publishers, Dordrecht, Netherlands.
17. Innis MA, Gelfand DH, Sninsky JJ, White TJ (1989) PCR Protocols: A Guide to Methods and Applications. Academic Press, New York.
18. Jefferson RA (1987) Assaying chimeric genes in plants: the GUS gene fusion system. Plant Mol Biol Rep 5: 387–405.
19. Jorgensen R, Snyder C, Jones JDG (1987) T-DNA is organized predominantly in inverted repeat structures in plants transformed with *Agrobacterium tumefaciens* C58 derivatives. Mol Gen Genet 207: 471–477.
20. Keohavong P, Thilly WG (1989) Fidelity of DNA polymerases in DNA amplification. Proc Natl Acad Sci USA 86: 9253–9257.
21. Klein TM, Wolf E, Wu R, Sanford JC (1987) High-velocity microprojectiles for delivering nucleic acids into living cells. Nature 327: 70–73.
22. Kwok S, Higuchi R (1989) Avoiding false positives with PCR. Nature 339: 237–238.

23. Maniatis T, Fritsch E, Sambrook J (1982) *Molecular Cloning: A Laboratory Manual.* Cold Spring Harbor Laboratory, Cold Spring Harbor, NY.
24. Mazodier O., Cossart P, Giraud E, Gasser F (1985) Completion of the nucleotide sequence of the central region of Tn5 confirms the presence of three resistance genes. Nucl Acids Res 13:195–205.
25. McClean P, Chee PP, Simental J, Held B, Drong RF, Slightom JL (1991) Susceptibility of dry bean (*Phaseolus vulgaris* L.) to *Agrobacterium* infections: transformation of cotyledonary and hypocotyl tissues. Plant Organ Tissue Cult 24: 131–138.
26. Mullis KB, Faloona F, Scharf SJ, Saiki RK, Horn GT, Erlich HA (1986) Specific enzymatic amplification of DNA *in vitro*: the polymerase chain reaction. Cold Spring Harbor Symp Quant Biol 51: 263–273.
27. Murray MG, Thompson WF (1980) Rapid isolation of high molecular weight plant DNA. Nucl Acids Res 8: 4321–4325.
28. Reynaerts A., De Block M, Hernalsteens J-P, Van Montagu M (1988) Selectable and screenable markers. In: Gelvin SB *et al.* (eds) Plant Molecular Biology Manual, Chap. A9. Kluwer Academic Publishers, Dordrecht, Netherlands.
29. Riggs CD, Bates GW (1986) Stable transformation of tobacco by electroporation: Evidence for plasmid concatenation. Proc Natl Acad Sci USA 83: 5602–5606.
30. Rogers SO, Bendich A (1988) Extraction of DNA from plant tissues. In: Gelvin SB *et al.* (eds) Plant Molecular Biology Manual, Chap. A6. Kluwer Academic Publishers, Dordrecht, Netherlands.
31. Saghai-Maroof MA, Soliman KM, Jorgensen RA, Allard RW (1984) Ribosomal DNA spacer-length polymorphisms in barley: Mendelian inheritance, chromosomal location, and population dynamics. Proc Natl Acad Sci USA 81: 8014–8018.
32. Saiki, RK, Scharf S, Faloona F, Mullis KB, Horn GT, Erlich HA, Arnheim N (1985) Enzymatic amplification of β-globin genomic sequences and restriction site analysis for diagnosis of sickle cell anemia. Science 230: 1350–1354.
33. Saiki RK, Gelfand DH, Stoffel S, Scharf S, Higuchi R, Horn GT, Mullis KB, Erlich HA (1988) Primer-directed enzymatic amplification of DNA with a thermostable DNA polymerase. Science 239: 487–491.
34. Saul MW, Shillito RD, Negrutiu I (1988) Direct DNA transfer to protoplasts with and without electroporation. In: Gelvin SB *et al.* (eds) Plant Molecular Biology Manual, Chap. A1. Kluwer Academic Publishers, Dordrecht, Netherlands.
35. Slighton JL (1991) Custom polymerase-chain reaction engineering of a plant expression vector. Gene 100; in press.
36. Slightom JL, Chee PP, Gonsalves D (1990) Field testing of cucumber plants which express the CMV coat protein gene: Field plot design to test natural infection presures. In: Nijkamp HJJ, van der Plas LHW, van Aartrijk J (eds) Progress in Plant Cellular and Molecular Biology, pp. 201–206. Kluwer Academic Publishers, Dordrecht, Netherlands.
37. Southern EM (1975) Detection of specific sequences among DNA fragments separated by gel electrophoresis. J Mol Biol 98: 503–517.
38. Talbot DR, Adang MJ, Slightom JL, Hall TC (1984) Size and organization of a multigene family encoding phaseolin, the major seed storage protein of *Phaseolus vulgaris* L. Mol Gen Genet 198: 42–49.
39. van den Elzen PJM, Towsend J, Lee KY, Bedbrook JR (1985) A chimaeric hygromycin resistance gene as a selectable marker in plant cells. Plant Mol Biol 5: 299–302.
40. Wang K., Herrera-Estrella L, Van Montagu M, Zambryski P (1984) Right 25 bp terminus sequence of the nopaline T-DNA is essential for and determines direction of DNA transfer from *Agrobacterium* to the plant genome. Cell 38: 455–462.

Index

Acetosyringone	A3/11
Acid Depurination	A15/17
Agarose	A/15/9, A15/15
LMP (low melting point)	A15/9, A15/19
Seaplaque (FMC)	A15/9
Aglycone	B14/3
Agrobacterium	A4/1, C3/1, C3/9, C3/20, C3/23, B14/9, B15/11
Biotypes	A4/1
Cointegration	A4/2
Growth Properties	A4/1
Introduction of Plasmid by Conjugation	A2/5, A4/2, A4/11
Introduction of Plasmid by Transformation	A3/7, A4/2, A4/9
-Mediated DNA Transfer	A2/1, A3/1, C3/1, C3/15
Plasmid Mini-prep	A3/9
Ri Plasmid (see also Plasmids)	A4/1
T-DNA	B15/11
Ti Plasmid (see also Plasmids)	A4/1
Total DNA Mini-prep	A2/5
TR2′ Promoter	B15/11
Alfalfa Mosaic Virus	B16/5
Anabaena spp	A12/4, 5, 12
Antisense RNA	A3/5
Aspergillus nidulans	A10/2
Avogadro's Number	A7/25
Bean cell-wall glycine-rich protein	B14/10
Benedicts Reagent	A4/7
Bertholletia excelsa	B16/43
Binary Vectors (see also Plasmids)	A3/1
β-mercaptoethanol	B14/27, B15/7, B15/9
BioGel P6	B14/13, B14/14
Biolystics (see Microprojectile Bombardment)	
Blot Hybridization Technique (see also Filter Hybridization)	C3/3
Bovine Serum Albumine (BSA)	A15/11, B14/19
5-bromo-4-chloro-3-indolyl β-D-glucuronide (X-Gluc)	A13/8, B14/4, B14/8, B14/10, B14/23

Buffers
Coupling Buffer B13/16
DNA Buffer B11/20
First Strand cDNA Buffer A14/11
Ø80 Buffer A8/15
Ø80 Storage Buffer A8/15
G50 Buffer C1/16
General Fixation Buffer B14/23
GR Buffer B5/9
Guanidine Buffer A14/5
GUS Extraction Buffer B14/3, B14/14
λ Dilution Buffer A7/49
Ligase Buffer A14/12, B16/15
M Buffer B3/26
MN Buffer B3/8
NT Buffer B3/8
NTE Buffer B9/22
NTS Buffer C1/4
PBS Buffer B3/10
Phosphate Buffer B14/23
Protoplast Fixation Buffer B14/23
Restriction Enzyme Buffers B13/16, A15/11
Second Strand cDNA Buffer A14/11
SM Buffer A14/16, A15/20
STC Buffer A10/9
STET Buffer B1/20
TAE Buffer C1/16
TBE Buffer B10/6, B11/12, B12/10, B16/14
TBST B13/16
TE Buffer A3/10, A7/19, A11/5, B2/10
TEA Buffer B2/10
TEB Buffer A14/20
TEM Buffer B16/10
TEN Buffer B16/15
TESE Buffer B12/10
X-Gluc Assay Buffer B14/24
Z Buffer B1/19, B15/6
cDNA Library Construction A7/1, A7/4, A14/6
Blunt Ending (of ds cDNA) A14/7
Directional Cloning (of cDNA) A14/14
First Strand Synthesis A7/3, A14/6

Gel Analysis (of cDNA) A14/12
Homopolymer Tailing A7/23
In vitro packaging and amplification of library A14/4
In vivo excision of pBluescript from λ Zap II (conversion of λ Library Into Phagemid Library) A14/16
Ligation of cDNA to Vector A14/9
Linker Addition A7/29, A14/7
Preparation of λ ZapII Vector A14/9
Protection of Sites A14/7
Screening A7/43
Second Strand Synthesis A7/11, A14/7
Size Fractionation (cDNA) A14/8
Transformation of *E. coli* A7/39
Cab Gene B10/1
Calothrix A12/4,5,12
Candida albicans A15/19
Carotenoids B14/24
Chloramphenicol B14/9
Chloramphenicol Acetyl Transferase (CAT) A9/3, B1/5, C3/1, B15/2
Chlorophyll B14/6
Chloroplast Isolation B5/5
Pea B5/16
Tobacco B5/5
Chromatin B2/1
Chromogenic Substrates B14/10
Cladosporium fulvum B14/10
Clonal Propagation B14/8
CaMV-CMV Promoter B16/8
CMV-C Coat Protein Gene B16/18
CNBR-Activated Sepharose 4B B13/14
Cofactors B14/3
Compression Zone (see Electrophoresis) A15/13
Complementation A14/1
β-Conglycinin B16/3, B16/35
Constitutive Plant Expression Cassette B16/5
Cos sites A15/19
Cosmid Vectors (see also Plasmids) A8/2
Cryostat Sections B9/3
Cryptic Glycosylation Sites B14/4
CTAB (Cetyltrimethylammonium Bromide) A6/1, C3/6
Cucumber mosaic virus C3/9

Cyclohexylammonium Salt B14/24
DEPC (Diethylpyrocarbonate) B9/7
Deprotection (Oligionucleotides) B16/44
5,5′-dibromo 4, 4′-dichloro-indigo B14/21
Dicotyledonous Species Plants B14/8
Differential Staining A15/7
Dimethylformamide B14/24
DMS (Dimethyl Sulfate) B12/1, B12/3
DMSO (Dimethyl Sulfoxide) A7/3
DNA
  Dephosphorylation A7/35
  Direct Transfer A1/1
  DNA Footprinting/*in vitro* B11/1, B11/13
  DNA Footprinting/*in vivo* B11/19, B12/1
  Electrotransfer B12/5
  End-labelling C1/5
  Isolation from
    Plant Tissues A6/1, B3/3, C1/11, C3/5
    Mitochondria A11/1
    Chloroplast DNA A15/3
  Naked Phage A15/19
  Non-linear DNA A15/3
  Partial Digestions A15/11
  Phage DNA from Mini-Lysates A8/11, A8/37
  Phage Lambda DNA A15/11
  Phage Lambda DNA concatemers A15/19
  Polymerase (Klenow Fragment) C3/2, B16/1, B16/43
  Preparation from DMS-Treated Cells B12/5
  Separation of DNA A15/13
  Strand Separation B10/6
DNA Methylation B3/1
  HPLC of Free Bases B3/5, B16/7
  Methylation Activity Analysis B3/23, B3/26
  Nearest Neighbor Analysis B3/7
  Restriction Enzyme Analysis B3/11, B3/14
DNase I B2/3, B11/1, B11/15
DTT (Dithiothreitol) A15/11, A15/19, B14/3
EDTA (Elhylenediaminetetraacetic Acid) A15/9, A15/11
Effector B14/1
Electrophoresis B12/7
  Agarose Gel B2/9, B5/8, B6/10, C2/7
  Countour Clamped Homogeneous Gel (CHEF) A15/1

Field Inversion Gel (FIGE)	A15/3, A15/15, A15/16
Glycerol Acrylamide Gel	B16/14
Polyacrylamide Gel (PAGE)	B1/16, B6/10
Programmable Autonomously Controlled (PACE)	A15/1
Pulsed-Field Gel (PFGE)	A15/1, A15/5, A15/19
Reversed Mobility Zone	A15/13
Reverse Orientation	A15/13
Rotating Gel (RGE)	A15/1
SDS-Page	B5/8, B5/19, B15/8, B15/9
Separation of Large Molecules	A15/15
Separation Range	A15/15
Separation Window	A15/14, A15/15
Separation Zones	A15/14
Time Ramps	A15/13, A15/14
Upper Compression Zone	A15/13
Electroporation	C3/15
Electroblotting	C3/15, C3/16, C3/17, C3/23
Endophytic Organisms	B14/9
Endoplasmic Reticulum (ER)	B14/4, B15/2
Enhancer Elements	B10/9
5-enolpyruvylshikimate-3-phosphate synthase	B14/10
Escherichia coli	
PLK-F′	A14/14
XL1-Blue	A14/14
LE392	A15/19
Escherichia coli β-glucuronidase (uidA)	B15/2, B15/3
E. coli β-galactosidase (*lac*Z)	B15/2–5, B15/7, B15/9–13
E. coli Polymerase (Klenow Fragment)	B16/43
Ethidium Bromide	A15/15
Etiolated Seedlings	B10/3
Exo-hydrolase	B14/3
Exophytic Organisms	B14/9
Ferredoxin	B5/2
Ferricyanide/ferrocyanide	B14/21, B14/23, B15/13
Filter Hybridization	B3/15, B12/9, C1/6, C2/15
Hybridization Solutions	A8/36, B3/21, B4/18, B9/22, B12/10, C1/16, C2/15
Northern Hybridization	B4/17
Southern Hybridization	C2/11
Fluoresceine Diacetate (FDA)	A15/7
Fluorescence Assays	B14/5, B14/16

Quenching B14/6
Microtitre Plate B14/16
Fluorochromes B14/27
Fluroescein β-D-glucuronide B14/29
Fluroescein Glucuronides B14/27
Fluorochrome Rhodamine B14/27
Fluorogenic Substrates B14/10, B14/27, B15/2
Resorufin B14/27
Resorufin β-D-glucuronide (ReG) B14/27
3-cyanoumbelliferyl β-D-glucuronide (CyUG) B14/27, B14/29
4-Trifluoromethylumbelliferyl β-D-glucuronide
(TFMUG) B14/27, B14/29
Fungi (see Transformation)
Gel Retardation Assay B11/1, B11/15
Gene Copy Number C3/19
Gene Expression B14/2
Gene Fusion B14/1–3, B14/8, B15/1
Gene Replacement A4/3
Gene Responder B14/1, B14/3
Genomic Sequencing B3/14, B12/1, B12/6
DNA Preparation B12/6
GLEAN (herbicide) A13/3
Glucaric Acid B14/17
Glucarolactone B14/17
Glucuronides B14/9
Gluteraldehyde B14/22
Glycine max B16/3
Glycosides B14/3
GUS (β-glucuronidase) A13/1, C3/1, B14/1, B15/3,
Assay A13/15, B14/9, B14/16
*gus*A gene B14/4
*gus*B gene B14/4, B14/11
*gus*C gene B14/4
gus operon B14/9, B14/28
*gus*R B14/4
Locus B14/4
Monomer Molecular Weight B14/4
Regulation B14/3, B15/1
Uid A gene B15/3
Helper Phage
R408 A14/17
Hexadecyltrimethylammonium bromide (CTAB) C3/5, C3/7

High Molecular Weight DNA A15/3, A15/9, A15/19
Histoprep B9/15
Hybrid Proteins B15/2
Hybridization (see Filter Hybridization, RNA Hybridization)
Immunocytochemistry
  Antibody Labelling B13/5, B13/19
  Antigen Fixation B13/9, B13/10, B13/11
  Antigen Modification B13/9
  Antibody Purification B13/14, B13/16
  Blocking B13/19
  Immunoaffinity Column Preparation B13/14
  Preparation of Protein A-gold B13/21
  Tissue Embedding B13/17
  Tissue Fixation B13/4, B13/17
  Tissue Sectioning B13/18
Initiator Codon B14/8
Intracellular Targeting B14/3
Introns B14/8
IPTG (Isopropyl-β-D-thiogalactopyranoside) A7/49
Ketolactose Test A4/7
Kodak OG X-ray Film B8/2
Kodak Technical Pan 2415 Film B9/27
Kodak XAR X-ray Film B8/2
λ Vectors
  Charon 4A A8/5
  Charon 34, 35 A8/3
  Charon 40 A8/3
  Charon Vectors A4/46
  EMBL3 A8/3
  EMBL4 A8/3
  λ Arms A8/4
  λgt 10 A7/3
  λgt 11 A7/3
  λ Wild Type A8/3
  λ-Zap II A7/3
Lac Operon B15/9
Lactic Acid B14/25
1-4 Lactone B14/17
Leaf Disc Transformation A5/1, A5/8, A9/1
Lignins B14/24
Light Induction B10/2, B10/4

Light-Regulated Sequences (LREs)	B10/8
Long-range Restriction Maps	A15/3
Lotus corniculatus	B14/10
Lucerne	B14/10
Luciferase Assay	B8/5
Luciferase Reactions (*in vivo*)	B8/7–9
Luciferin	B8/2, B8/5, B8/7
Luminographs	B8/2, B8/10
Luteolin-triglucuronide-specific β-Glucuronidase	B14/8
Lysis Mix	A15/9, A15/10, A15/19
Lyticase	A15/20
Maize (Black Maxicam Sweet, BMS)	A13/2, A13/5
Male Gametophytic Tissue	B14/8
Markers (See Transformation Markers)	
Media	
A Medium	A1/6–8
B Medium	A13/13
B5 Medium	A5/6
B5 Vitamins	A13/13
Basic Medium	C1/10
C Medium	A13/13
D Medium	A13/13
E Medium	A13/13
F Medium	A13/13
H Medium	A1/6–8
K3 Medium	A1/6–8
Lactose Medium	A4/7
LB Medium	A2/7, A4/4, A5/6
LS Medium	A1/6–8
Minimal A Medium	A14/20
MM (Minimal Medium)	A4/5
Modified Barth's Medium	B7/5
MS Medium	A3/5
MS Macronutrients	A13/12
MS Micronutrients	A13/12
MS Rooting Medium	A5/6
MS Salt Base	B10/4
MS Selection Medium	A5/6
MS Suspension Medium	A5/6
MS Vitamins	A13/12
MS104 Medium	A5/6
MS0 Medium	A5/6

Murashige and Skoog Medium	A15/6
NB Medium	A4/4
NZ Top Agar	A8/15, A8/36
NZC Bottom Agar	A8/15
NZC Medium	A8/15, A8/36
NZY	A14/16
NZY Top Agar	A14/16
O-R2 Medium	B7/5
Regeneration Medium	A10/9
SB Medium	A14/18
SOB Medium	A14/19
SOC Medium	A7/41
T Medium	A1/6-8
TB Medium	A14/16
TY Medium	A4/4
YEP Medium	A3/5
YMB Medium	A4/4
YPD Medium	A15/19, A15/20
YT Medium	A14/18
Methylcytosine (5-)	B3/1
Methylumbelliferone 4-(MU)	A13/15, B14/6, B14/15–17, B15/9
Methyl-umbelliferyl-β-D-glucuronide (MUG)	A13/15, B14/4–6, B14/15–17, B14/19, B14/27–29, B15/3, B15/7, B15/8, B15/10
Methylase	A14/17
Mesophyll Cells	B14/7
Micrococcal Nuclease	B2/5
Microinjection of Oocytes	B7/11, B7/14
Micromanipulator	B7/7
Micropipette (Construction and Maintenance)	B7/9, B7/18–19
Microprojectile	A13/1, A13/7, C3/1, C3/15
Minimum Mobility Zone (see Electrophoresis)	A15/13
Mitochondria, isolation of	A11/3
Monitoring Gene Activity *in vivo*	B14/2
Monocotyledonous Plants	B14/8
Multigene Families	B14/2
Mung Bean Nuclease	B16/1
Mycoplasma	A15/2
N-linked Glycosylation	B15/2
Naphthol ASBI-glucuronide (NAG)	B14/28, B14/29

Neurospora crassa	A10/2, A15/19
Neomycin Phosphotransferase (NPTII)	A9/7, B1/9, B4/7, B5/9, B15/2
Nick Translation	B3/15, C2/1,13,16
Nicotiana tobaccum	A13/5
Nopaline Synthase (NOS)	A2/2, C3/9, B15/2
Assay	A13/15
Northern Blotting (see also Filter Hybridization)	B4/17
NOS-NPT II' – NOS Gene	A2/2, A3/3, C3/9, C3/23–25
Nostoc	A12/4, A12/12
NPT II Assay	A13/19, B1/15, B1/9
Nuclear Extracts Preparation	B11/9
Nuclei Isolation	B2/3, B11/7
Nylon Filter	C3/16, C3/23, C3/25
Octopine Synthase (OCS)	B15/2
Oligonucleotide Primers	C3/3, C3/9
ONP (Orthonitrophenol)	B15/6, B15/7
ONPG (Orthonitrophenol)-D-Galactopyranoside)	B15/5, B15/6
OPC Purification Cartridge	B16/7, B16/43
Patatin-GUS fusions (see also Gene Fusion)	B14/10
Papaya Ringspot Virus (PRV)	B16/22
Papaya Ringspot Virus Coat Protein Genes	C3/9
PEG precipitation	A15/19
Percoll Gradient	A11/3, B5/5, B5/14
Petunia	B14/10
PFGE (see also Electrophoresis)	A15/3, A15/14, A15/17, A15/19
Phage Lambda DNA (see also λ Vectors)	A15/11
Phaseolin Expression Cassette	B16/28
Phenophthalein β-D-Glucuronide	B14/28
Phenyl Methyl Sulphonyl Fluoride (PMSF)	A15/11, B14/13
Phenylalanine Ammonia Lyase	B14/10
Phaseolus vulgaris	C3/5, B14/10, B16/3
Photinus pyralis	B8/1
Phosphinothricin Acetyl Transferase (BAR)	B15/2
Phytochrome-induced Gene Expression	B4/25, B4/27
Plant Promoters	
35S-CAMV	A3/5, A13/3, B14/7, B14/8, B14/10, B15/6
Light-regulated	B10/1
Plant Tissue Transformation	A3/13

Plasmids

pABD1 A1/1
pAGV40 B1/3
pAL4404 A3/1
pAT153 A7/23
pB0542 A3/1
pBluescript B9/7
pBR322 A7/2, A7/23, A7/33, A7/35
pDH51 B16/5
pDO432 B8/7
pDR100 B8/7
pDR101 B8/7
pDR102 B8/7
pDR103 B8/7
pDS4101 A12/17
pG428 A12/17
pGA482 A3/3, B16/7
pGA482GG B16/23
pGA492 A3/4, B1/3
pGA515 A3/5
pGA553 Series A3/4
pGA580 A3/4
pGA581 A3/4
pGA582 A3/4
pGA583 A3/4
pGA642 A3/5
pGA643 A3/5
pGEM-5Zf(+) B9/7
pGEM-7Zf(+) B9/7
pJ21LA B1/3
pJ215 B1/3
pKC7 B1/5
pMC1871 B1/4
pMON200 A2/1
pMON9749 A13/9
PMON9909 A13/9
pPHIJI A4/3
pPvPH3.8β B16/29
pRK2 A3/15
pRK290 A4/3
pRK2013 A2/2
pRL518 A12/17

pRL528 A12/17
pRL530 A12/17
pRL543 A12/17
pRL566 A12/17
pSKS107 B1/4
pSP64 B5/10, B7/1
pSP65 B5/10, B7/1
pTiA6 A2/2
pTiB6S3-SE A2/2
pTiT37 A2/2
pTJS75 A3/1
pUC 8 A1/1, A9/11
pUC1318 B16/23
pUC1813 B16/23
pUC1813cpCMV19 B16/5
pUC18cpexp.-17 B16/20
pUC18cpWMVII-5 B16/22, B16/25
pUC18pvPTexp7-7 B16/30
pUC18pvPSPTexpD22 B16/30, B16/34
pUC NKI B1/3–5
pUC Series A7/23, A7/35
pvPSPTexp B16/27, B16/33
pWC12 B5/10
pWMVII-3.2 B16/22
Ri Plasmid A3/13
Ti Plasmid A4/1
Plastocyanin B5/2
PMSF (Phenyl methyl sulfonide fluoride) A15/11, B14/13
p-nitrophenyl β-D-glucuronide (PNPG) B14/4, B14/4, B14/19, B14/28, B14/29
Poly-D-lysine B9/16
Polyacrylamide Gels B1/16
Polyclar AT B14/13
Polyadenylation Signals B14/8
Polyphenolics B14/13, B14/24
Polymerase Chain Reaction (PCR) C3/2–4, C3/9–13, C3/15 B16/1
Custom PCR (CPCR) B16/2
CPCR DNA Product B16/9
CPCR Engineering B16/6
Inverse PCR A15/2, C3/4
PCR Buffer C3/12

PCR-Generated Errors B16/2
PCR Template B16/1
Poly-U Sepharose B4/15
Polyvinylpyrrolidone B15/7
Post-transcriptional Modification B14/4
Post-transcriptional Mechanisms of Regulation B15/2
Primer Annealing C3/2
Primer Extension Method C1/5
Probe Labelling
3' End-labelling/Klenow B11/11
5' End-labelling/Polynucleotide Kinase B11/13, C1/5
M13 Primer Extension B3/19, B12/9
Nick Translation B3/15, C2/13
Radio-labeled RNA B7/4, B9/9
Protein-A-gold B13/21
Protein-DNA Interactions B12/1
Protein Labelling (in oocytes) B7/11
Proteinase K B9/8, B9/19, A15/9, A15/10
Protoplast
Fixation Solution B14/23
Isolation A1/9
Preparation A15/5
Protoplast Forming Enzyme A15/20
Non-viable A15/7
Regeneration A1/13
Transformation Selection A1/13, A9/1
Transformation/with Electroporation A1/9
Transformation/without Electroporation A1/11
rbcS Gene B10/1
Regenerated Plants A13/11
R1 Progeny A13/11
Reporter Genes (see also Transformation Markers) B1/1, B5/1, B14/1, B14/2, B15/1, B15/2
Restriction Enzyme Digestion C2/5, A15/11
Restriction Enzymes (Types I & II) C2/1
Rare Cutter A15/3
Naturel Partials A15/3
Restriction Fragment Length Polymorphism (RFLP) C2/1
RFLP Maps A15/2
Rhizobium genes B14/10

RNA
  Antisense A3/5
  Blot analysis B4/7, B4/21
  Denaturation A7/5
  mRNA Isolation A14/3, B6/3
  mRNA Synthesis B5/10
  Poly(A)$^+$ RNA Preparation B4/15
  Polysomal RNA Isolation B6/7
  Total RNA Preparation B4/11, B4/13, C1/3
  Transcription, *in vitro* B5/10, B7/1, B7/3, C1/14
  Translation, *in vitro* B5/13
  Translation, *in vivo* B6/10
RNA Hybridization (*in situ*) B9/1, B9/28–30
  Embedding/Sectioning of Tissue B9/1
  Synthesis of RNA Transcripts B9/7
  Analysis of RNA Transcripts B9/13
  Pre-hybridization B9/3, B9/19
  Hybridization B9/3, B9/21
  Autoradiography B9/5, B9/25
  Immunocytochemistry B9/4
Root Nodule B14/10
S1-Nuclease Protection B10/6, B16/1
  Assay B4/23
Saccharic Acid B14/17
Saccharolactone B14/17
Saccharomyces cerevisiae A15/19
Sarcosyl B14/13
Schizosaccharomyces pombe A15/19
Selectirve Turnover of Proteins B14/3
Sepharose CL6B B14/13, B14/14, B14/19
Silencers B10/8
Sodium Dodecyl Sulphate A15/9
Sodiummetabisulphite B14/24
Sodium N-lauroylsarcosine A15/9
Solutions
  Antibiotic Stock Solutions A2/7
  BLOTTO B13/21
  CPW-salts Solution A15/6
  Denaturation Solution C3/18
  Denhardt's Solution A7/45, B4/19, C2/15, C3/18, C3/19
  DnD Solution A14/20

EB Solution B11/16
Fixative B13/18
INTP C3/12
λ Infection Salts A7/49
Lysis Mix A15/10, A15/9, A15/19
Neutralization Solution C3/18
PBS Solution B9/17
PLC Solution C1/4
SSC Solution A7/45, A8/36, B3/16, B4/19, C1/16, C2/15
SSPE Solution B9/20
Sucrose Solution A15/5, A15/6
W5 Solution A1/12

Southern Blotting C2/9, C2/12, A15/10, A15/17
Genomic Blot (see also Filter Hybridization) C1/11, C3/1–4, C3/15, C3/18–23

Spectrophotometric Methods B14/5
Spermidine A15/11
Strong Background Signal A15/20
Synechochococcus spp A12/2, A12/4, A12/5, A12/12
Synechocystis, PCC 6803 A12/2, A12/4, A12/5, A12/12
Taq Polymerase B16/2
T-4 Polymerase B16/1
Target Gene C3/15
Target Sequences B14/9
T-DNA (see Ti Plasmid) A2/1
Temperature Gradients in FIGE (see also Electrophoresis) A15/15
Thermal Inactivation B14/4
Thermostable DNA polymerase (*Taq*) C3/2, C3/3, C3/10
Thermus aquaticus C3/2
Ti Plasmid (see also Plasmids) A4/1
Cointegrating Vectors A2/1
T-DNA A2/1
T-DNA Borders A3/3
T-DNA Cloning Sites A3/3

Tissue (see also Immunochemistry)
Embedding B13/17

Fixation B13/4, B13/17, B13/19
Sectioning B13/18, B14/22
Tissue Fixation (see also Immunochemistry)
Glutaraldehyde B13/14
Formaldehyde A13/14
Removal of Osmium with Periodate B13/19
Transcriptional Fusions (see also Gene Fusion) B15/1
Transformation
Cell Suspension A3/11
Cyanobacteria A12/2, A12/11
E. coli A7/39
E. coli Auxotrophs (complementation) A7/39
Fungi A10/1
Leaf Disc A5/1, A5/8, A9/1
Plant Cell, Intact A13/1
Plant Tissue A3/13
Protoplast/with Electroporation A1/9
Protoplast/without Electroporation A1/11
Transformation of Fungi A10/1
Aspergillus nidulans A10/2
β-Glucuronidase A10/4, B14/17
Filamentous Fungi A10/2
Hygromycin B Selection A10/9
Lithium Mediated Transformation A10/5
Neurospora crassa A10/2
Plant Pathogenic Fungi A10/2
Transformation by Electroporation A10/5
Transformation of Protoplasts A10/4, A10/7
Transformation with a Particle Gun A10/5
Yeast A10/1
Transformation Methods B15/1
Transformation Markers A9/1, A9/16, B8/1
β-Galactosidase B1/6, B1/10, B1/19, B15/2–3, B15/5–12
β-Glucuronidase (GUS) A13/1, A13/8, B14/17
Bacterial Luciferase (luxAB) B8/2, B15/2
Callus Induction Assay A9/5
CAT (Chloramphenicol Acetyl Transferase) A9/3, B1/5, C3/1, B15/2
CAT Assay A9/12, B1/8, B1/13, B4/3
Firefly Luciferase (luc) B8/1, B15/2
Hormone Biosynthetic Genes A9/1
Hygromycin Phosphotransferase (hpt) A9/1, C3/1

NPT II (Neomycin Phosphotransferase) A9/7, B1/9, B1/15, B4/7, B5/9, C3/1, B15/2
NPT II Assay A9/1, B1/13, B1/5, B5/1
Opine Synthase Assay A9/13, B1/11
Opine Synthesis Genes A9/3
Screenable A9/3, B1/1
Segregation of Transferred Genes A9/11, C1/9
Selectable A9/1
Transgenic Plants B14/3, B15/13
Translational Efficiency B14/8
Translational Fusions (see also Protein/Gene Fusion) B15/1, B15/2
Translational Initiation B14/3
Translocation of Proteins Across Membranes B14/3
Transit Peptide B5/1
4-Trifluoromethylumbelliferyl β-D-glucuronide (TFMUG) B14/27, B14/28
Triparental Mating A2/5, A12/15
Trityl-bearing Oligonucleotides B16/12
Ultraviolet Microscopy A15/7
UV Irradiation A15/17
Velocity Sedimentation Gradients A8/17
Viability Determination A15/7
Differential Staining A15/7
Fluoresceine Diacetate (FDA) A15/7
Non-viable Protoplasts A15/7
Virulence Functions (vir) A3/1
Whole Cell Extract Preparation A11/15
WMVII Coat Protein Gene B16/21
X-gal A7/49, B15/3, B15/4, B15/9, B15/10, B15/13
X-glu (5-Bromo-4-chloro-3-indolylglucuronide) A13/8
Xenobiotic B14/3
Xenopus Oocytes (preparation) B7/8
YAC Vectors A15/2
Yeast Chromosomes A15/1, A15/19, A15/19
Yeast Artificial Chromosome (YAC) A15/2
Zucchini Yellow Mosaic Virus (ZYMV) B16/22
Zymolyase-100T A15/19, A15/20

GPSR Compliance

The European Union's (EU) General Product Safety Regulation (GPSR) is a set of rules that requires consumer products to be safe and our obligations to ensure this.

If you have any concerns about our products, you can contact us on ProductSafety@springernature.com

In case Publisher is established outside the EU, the EU authorized representative is:

Springer Nature Customer Service Center GmbH
Europaplatz 3
69115 Heidelberg, Germany

Batch number: 10620141

Printed by Printforce, the Netherlands